COLOUR & MONO TELEVISION
for
CITY & GUILDS AND TEC COURSES
Volume 1
Television Reception: Principles & Circuits

COLOUR & MONO TELEVISION
for
CITY & GUILDS AND TEC COURSES

Volume 1

Television Reception:
Principles and Circuits

K. J. BOHLMAN

T. Eng., F.S.E.R.T., A.M.Inst.E.
Senior Lecturer in Radio and Television
Lincoln College of Technology

LONDON
NORMAN PRICE (PUBLISHERS) LTD

NORMAN PRICE (PUBLISHERS) LTD
17 TOTTENHAM COURT ROAD, LONDON W1P 9DP

ISBN 0 85380 135 5

Printed in Great Britain by
Biddles Ltd, Guildford, Surrey

PREFACE

This is the first of a new series written specially for mechanics and technicians who are concerned with the reception of television signals.

Volume 1 deals with colour theory, television principles and 625-line receiver circuit techniques up to and including the video and audio output stages. Circuits employing discrete components are used to illustrate basic principles and where applicable integrated circuit versions are also described.

In dealing with the propagation of the television signal and receiving aerials the effects in Bands 1 and 3 have been discussed, no decision having been reached as to the future use of these bands when the 405-line transmissions are phased out in 1982.

The text has been written with the needs in mind of students following City & Guilds courses and those on TEC courses at technical level as well as to others who are concerned with the maintenance of television receivers.

Succeeding volumes will deal with mono and colour display tubes, sync. separation, timebases, a.g.c., convergence, colour decoder and digital applications in television.

CONTENTS

CONTENTS

CHAPTER 1

LIGHT AND HUMAN VISION

INTRODUCTION

TELEVISION is the art of reproducing at some distant point an instantaneous visual image of a scene using a system of telecommunications. A SCENE can be taken to mean any area or object that is capable of sending out light. Light from most scenes reaches our eyes by the mechanism of reflection. As the world about us is abundant with colour, the light emanating from everyday scenes is, of course, coloured. Thus an IDEAL television system will reproduce a scene at a remote point in the same way as we would receive the scene by direct viewing, *i.e.* in full colour detail and provide a sensation of depth. Until 1967, television transmissions in this country were confined to the production of black-and-white (monochrome) images using a cathode-ray tube screen emitting white light. Over the years technical improvement has resulted in a brighter, sharper and more stable black and-white-image. However, the reproduction of a coloured scene in monochrome is an unnatural process and subject to standards which to some extent are a matter of opinion. With the advent of colour television transmissions, the missing dimension of 'colour' has now been included thus providing the viewer with added realism and bringing the television system closer to the ideal.

Monochrome t.v. engineers who were concerned only with the servicing of receiver equipment did not consider it necessary to study the subject of 'light' and indeed rarely was the subject mentioned in books dealing with the principles of monochrome receiving equipment. However, to obtain a clear understanding of colour television a knowledge of certain aspects of light is necessary. The important properties of light and the human seeing senses in so far as they affect the subject of television will be considered in this chapter.

ELECTROMAGNETIC WAVES

Radiant heat from an electric fire travels through space in the form of electromagnetic waves. In contrast to the conduction and convection of heat no medium is required to propagate radiant heat, which may pass through a vacuum. An electromagnetic wave consists of an oscillating electric field together with an associated magnetic field, coexisting at right angles to one another, Fig. 1.1. The movement of these two fields through space constitutes an electromagnetic wave.

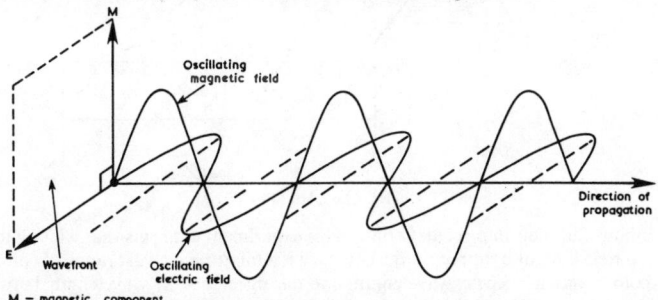

M = magnetic component
E = electric component

FIG. 1.1   REPRESENTATION OF AN ELECTROMAGNETIC WAVE

Light rays, radio waves, X-rays and cosmic rays, like radiant heat, travel through space in the form of electromagnetic waves and with high velocity (3×10^8 metres per

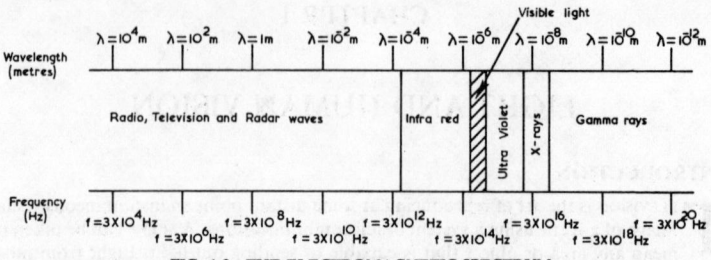

FIG. 1.2 THE ELECTROMAGNETIC SPECTRUM

second in a vacuum). It is evident, therefore, that these different types of radiation are all essentially of the same form. What then distinguishes one form of energy from another? The properties of the different types of electromagnetic energy are determined by their frequency (or wavelength), see Fig. 1.2. For convenience, in this diagram the electromagnetic spectrum is divided into sections but the sections have no precise boundaries since the behaviour of an e.m. wave does not change sharply at given frequencies. The wavelength (λ) of an e.m. wave may be found from

$$\lambda = \frac{v}{f} \text{ metres}$$

where f = the frequency (Hz), v = the velocity of e.m. waves (3×10^8 m/s) and λ = the wavelength (metres).

Visible light occupies a very small section of the e.m. spectrum with wavelengths in the approximate range of 380–780 nano metres (1 nano metre = 10^{-9} metre). When all the frequencies present in the visible section reach the eye simultaneously we see 'white light'. For the present we may regard the human eye as a 'window' for the brain in the e.m. spectrum. The human 'seeing senses' behave as a high Q tuned circuit, selecting the small visible section but rejecting the wavelengths on either side of this section.

SPECTRAL COLOURS

When a beam of white light as in Fig. 1.3 falls on a glass prism, the ray that emerges is no longer a beam of white light but a divergent beam containing all of the colours of

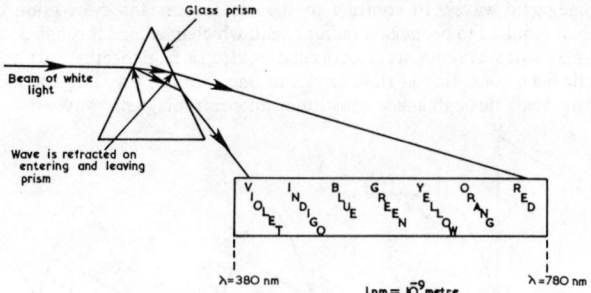

FIG. 1.3 THE COLOUR SPECTRUM

the rainbow and their intermediate tints. This experiment confirms that white light is not the purest kind of light but on the contrary is a mixture of a vast range of colours. Each colour has a specific wavelength but the difference in wavelength between adjacent colours in the spectrum is so small that the vast range of colours present are not distinguishable by the human senses. Instead, we see a gradual blending of the numerous colour radiations which produces a graduation of distinct hues called SPECTRAL COLOURS. As an aid to memory, the initial letters of the spectral colours spell the name ROY G. BIV.

Divergence of the beam is the result of refraction at the incident and exit faces of the glass prism. The amount of refraction or 'bending' of the beam is greater for the shorter wavelengths, *i.e.* more at the 'blue' end of the spectrum than the 'red' end.

THE CHARACTERISTICS OF A COLOUR

There are many words in common use describing colour. Apart from the colours which have their own names, *e.g.* green, red, yellow, orange, etc., there are many qualifying words used such as *vivid* red, *dark* blue, *pale* yellow, and *bright* orange etc. For television purposes these descriptions of colour are not precise enough, How, for example, can we distinguish between two slightly different shades of pale yellow? What really is the essential difference? In television, it is necessary to have an accurate way of specifying a colour so that its particular subtlety may be converted into an equivalent electrical voltage for subsequent reproduction of the original colour.

Any colour may be fully described by three characteristics: (a) Hue, (b) Saturation; and (c) Luminance.

(a) Hue

Hue is the quality of a colour most noticeable to the human senses. It is the preferred term for colour, *e.g.* when white light falls on to a red object the light reflected has a red hue. Similarly, if the object is green the light reflected has a green hue, Fig. 1.4(a) and (b). Thus hue depends upon the dominant wavelength of the light energy and can be specified by quoting its position in the colour spectrum.

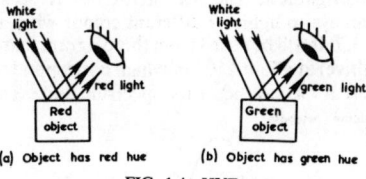

(a) Object has red hue   (b) Object has green hue

FIG. 1.4   HUE

(b) Saturation

Saturation is a quality of a colour which defines its depth. The saturation of a colour depends upon its dilution with white light. A colour which is highly saturated has most of its energy concentrated about a predominant wavelength, see Fig. 1.5, which shows a saturated red hue. When white light is added to a saturated red hue a

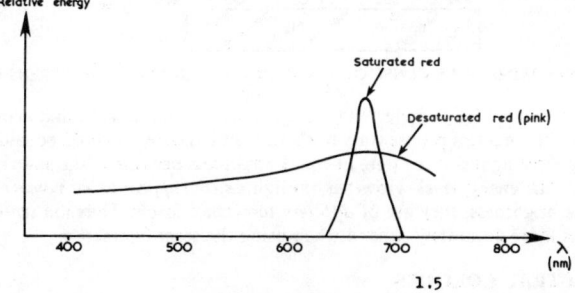

1.5

FIG. 1.5   SATURATION

paler shade of red is produced which we call pink. Pink has the *same hue* as red because the red wavelengths are predominant. The greater the amount of white light that is added, the paler or more *desaturated* the hue becomes. Desaturated hues are commonly called 'pale shades' or 'pastel colours'. Most of the colours that we see about us are desaturated to various extents.

(c) Luminance

This quality describes the brightness of a colour *as assessed by the human eye.* Consider Fig. 1.6 which shows three equal energy coloured light sources throwing out separate red, blue and green beams. Now, to an observer moving along the line *AB*, the

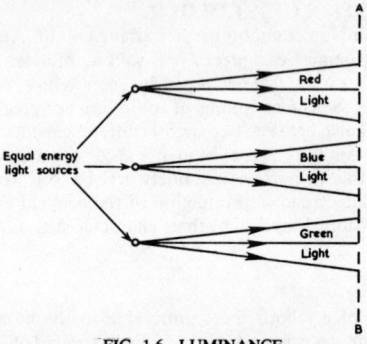

FIG. 1.6   LUMINANCE

colours will not appear to be of the same brightness even though the beams are projected with the same energy. It will be noticed that the green light will appear the brightest of the three, with the red the next brightest and the blue producing the smallest sensation of brightness. The reason for this is because of the non-linear response of the human eye to lights of different colour when they are radiated with equal energy, see Fig. 1.7. It will be noted from this diagram that the response of the eye is greatest (most sensitive) at about 550 nm which corresponds to a yellow-green hue. The response drops off at the red end of the spectrum and even more at the blue end.

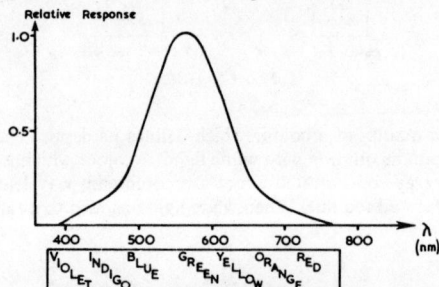

FIG. 1.7   APPROXIMATE RESPONSE OF STANDARD HUMAN EYE TO DIFFERENT COLOUR RADIATIONS

The term 'luminance' is preferred to 'brightness' which is a word that in everyday speech describes in a less precise way 'brilliance' and 'dazzle'. It could be said that in Fig. 1.6 the three light sources were of equal brightness because of the association of 'brightness' with 'energy level'. However, the lights *do not appear to the human eye* to be of the same brightness; they are of *different luminance* levels. Thus the *sensation of brightness* is more accurately defined when using the term luminance.

NON-SPECTRAL COLOURS

(1) Purple (Magenta)

This colour may be produced by mixing blue and red light. Since red and blue appear at opposite ends of the colour spectrum they do not merge with one another.

(2) Brown

This is not a true colour but is the sensation of viewing dirty orange or dirty yellow in contrast with a brighter area.

(3) Black

Black, of course, is the absence of light energy at all wavelengths. However, dull colours may appear black when viewed near to bright light. This is because of the limited range of brightness levels that the eye can register at the same time. A particular example of this illusion is a monochrome receiver screen which when unenergised has a dull grey or green appearance. However, when the screen is emitting white light, the non-energised areas of the screen appear black in contrast to the bright white areas so creating a 'black-and-white' picture.

WHITE LIGHT

When all the visible light radiations reach the eye simultaneously we see white light. There are, however, many sources of so-called 'white light' but these have quite different spectral energy distributions, see Fig. 1.8. Some white light sources give out more energy towards the red end of the spectrum and are referred to as 'warm whites'.

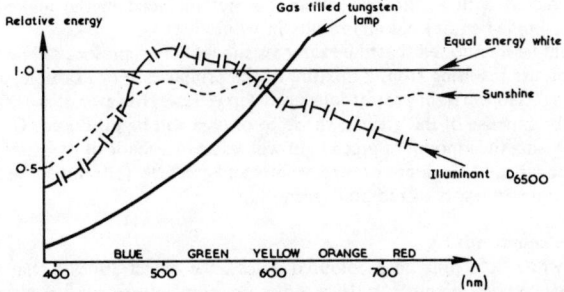

FIG. 1.8 SPECTRAL RESPONSE OF VARIOUS 'WHITE LIGHT' SOURCES

On the other hand some white light sources emit more energy at the blue end of the spectrum producing a 'cold white'.

Because the type of white light source used in the television studio will have a pronounced effect on the actual colours 'seen' by the camera, it is necessary to adopt a standard illuminant. For colour television the standard white Illuminant D_{6500} was chosen since it matches standard daylight more closely than other standard illuminants. By using a standard white light more consistent colour pictures are possible. In setting up the white light emitted from the screen of a colour television receiver, the light output of the three phosphors is adjusted to obtain a match with Illuminant D_{6500}. The type of white light emitted from the screen of a monochrome display tube depends upon the chemical composition of the screen coating. Some monochrome display tubes emit light which matches the standard illuminant to satisfy the requirements of television studios.

Equal energy white is the type of white light that would be produced if all the visible light radiations were of the same energy level. This type of white light is not a practical one, but is used as a standard in simplifying colour theory studies.

COLOUR MIXING

The principle of colour printing, colour television and colour photography is based on the mixing of usually three colours to obtain all the colours in the original. In colour television, the three colours (referred to as 'primary colours') are red (615 nm), green (532 nm), blue (470 nm). By suitable mixing of the three primaries it is possible to produce nearly all the colours that occur in nature.

There are two ways of mixing colours: by mixing coloured lights (the additive system) used in colour television; or by mixing coloured pigments, *e.g.* inks dyes and paints (the subtractive system).

Additive Colour Mixing

If two or three of the t.v. primary colours are projected from lamps on to a white screen so that they partly overlap, the colour that we see in the overlap area is the result of the additive mixture of the original primaries. Figs. 1.9* and 1.10* show the result when various coloured lights are additively mixed.

Note that when all the t.v. primaries are additively mixed (in suitable proportions) the result is white. In addition, white is also produced when the following coloured lights are additively mixed:

$$Blue + Yellow = White$$
$$Green + Magenta = White$$
$$Red + Cyan = White$$

Magenta, cyan and yellow are known as 'complementary colours'. Magenta is complementary to green, cyan is complementary to red and yellow is complementary to blue. It follows, therefore, that a complementary colour is one which when additively mixed with a primary colour (*i.e.* not included in the make-up of the particular complementary colour) results in white.

It should be mentioned that the examples shown here represent only a few of the possible colours resulting from a mixture of the primaries. For example, adding red and green lights in the right proportions results in yellow. However, if more red light is added at the expense of the green a shade of orange will be produced. On the other hand, increasing the amount of green light will result in a shade of greenish-yellow. In fact a whole range of shades of orange, yellow and greenish-yellow may be created by mixing discrete amounts of red and green light.

Subtractive colour mixing

When white light falls on a coloured object, the object appears that particular colour because the pigment reflects only its own colour whilst absorbing or SUBTRACTING the other radiations of the white light. We have seen that white light can be considered to be composed of red, blue and green lights. Thus if white light falls on a red object, the red light is reflected but the green and blue components are subtracted from the incident white light, Fig. 1.11(a). If the object is yellow, the pigment will reflect the red and green components but absorb the blue component, Fig. 1.11(b). When, say, magenta light falls on a blue object as in Fig. 1.11 (c), the pigment reflects the blue component of the incident magenta light but absorbs the red component, thus the object still appears blue.

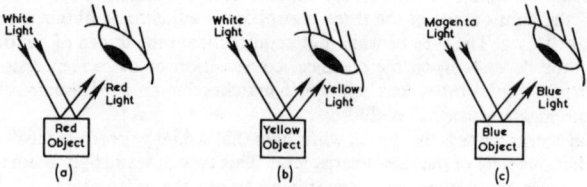

FIG. 1.11 PIGMENTS

It is well known that coloured objects appear to change their colour when seen under different colour light sources, *e.g.* under street lighting (sodium lamps). In Fig. 1.12(a) the yellow object appears red when illuminated with red light (note that the yellow pigment is capable of reflecting red and green light). The green object of diagram (b) appears black, *i.e.* no light is reflected when illuminated by blue light since all colours except green will be absorbed by the green pigment.

When pigments are mixed the resultant colour is that which is *common* to the

*In the fold-out colour section at the front of the book

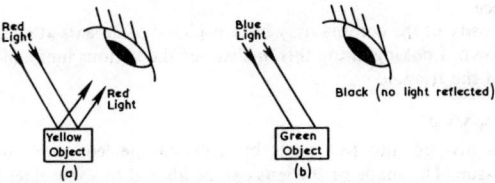

(a) (b)

FIG. 1.12 CHANGE OF COLOUR OF PIGMENT WHEN VIEWED UNDER CERTAIN COLOUR LIGHTS

original colours of the mixture. For example, a mixture of yellow and magenta paints produces red.

$$\text{YELLOW} + \text{MAGENTA} = (red + \text{green}) + (red + blue).$$

A mixture of cyan and magenta inks will result in blue since this is the common colour

$$\text{CYAN} + \text{MAGENTA} = (\text{green} + blue) + (blue + \text{red}).$$

When cyan, magenta and yellow pigments are mixed the result is black since there is no common colour. This method of mixing is used by artists. The colours yellow, cyan and magenta are referred to as the 'artists' primary colours' but are commonly called yellow, blue and red. Subtractive mixing of the artists' primaries is shown in Fig. 1.13.*
The same principle applies to the mixing of colour filters. Consider Fig. 1.14* which shows white light incident upon a combination of yellow and magenta filters. The yellow filter will pass red and green light whereas the magenta filter will pass red and blue light. Thus, when the two filters are in combination or 'mixed' the light passing through the combination is that which is common, i.e. red light.
Most people are more familiar with the mixing of pigments than coloured lights; the subtractive system has been described here simply to point out the essential differences between the two methods of mixing colours. Subtractive mixing, of course, is not used in colour television.

THE COLOUR TRIANGLE

The characteristics of a colour and the mixing of coloured lights having now been considered, these features will now be combined in a useful diagram called the 'colour triangle', Fig. 1.15.*
In this diagram, primary coloured light sources of red, green and blue are assumed to be placed in the corners of an equilateral triangle and the beams of light directed inwards. The following points may be seen from the diagram:

(a) Additive Mixing

Yellow, magenta and cyan will be produced along the sides of the triangle due to the additive mixture of pairs of primary colours. Assuming that the primary light sources are of suitable relative intensity, white will be produced at the centre of the triangle (W).

(b) Hue

As the eye travels around the circumference of the circle in the triangle the colour of the light changes, i.e. the HUE is seen to vary from red, orange, yellow on to cyan, etc.

(c) Saturation

As the saturation of a colour depends upon its dilution by white light, saturation may be defined by considering a line such as RW. The nearer one moves along this line towards W the more desaturated or paler the hue becomes because more white is added to the red hue. At W the hue is fully desaturated and at R is fully saturated. Note that all points along this line have the same hue, i.e. red.

*In the fold-out colour section at the front of the book

(d) Luminance

The luminosity of the colours may be considered as an axis at right angles to the triangle as shown. Looking along this line we see the various luminance levels of the colours within the triangle.

THE HUMAN EYE

The eye is divided into two parts by a crystalline lens made up of layers of transparent tissue. The shape of the lens can be altered to some extent by the ciliary muscles, see Fig. 1.16. Immediately in front of the lens is the iris. This is the coloured portion of the eye that can be seen through the transparent bulge of the cornea. The iris

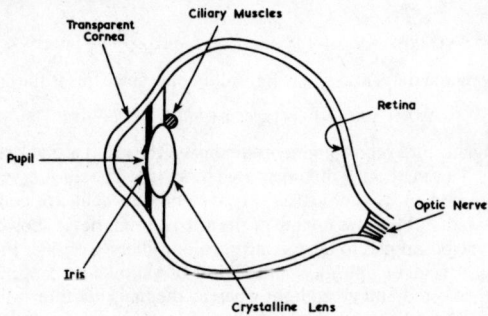

FIG. 1.16   THE HUMAN EYE

is an opaque screen with a small hole in the centre called the pupil. The size of the pupil is controlled involuntarily by the amount of light falling on the eye. At the back of the eye is the retina. In order to see an object clearly it must be brought to a focus at the centre of this surface (called the yellow spot). The retina is composed of a number of layers and at the back are the 'light sensitive elements'. There are two types of elements called rods and cones (on account of their shape when viewed under a microscope). When light falls on these receptors, electrical impulses are fed to the brain *via* the optic nerve where the image is registered.

In normal vision, in good light, the image is brought to a focus on a very small region at the centre of the retina. This small area is found to contain cones only, thus these receptors must be responsible for detailed and colour vision. Away from the centre of the retina the cones get less and less in proportion to the rods. Whilst the rods give poor detailed vision and little colour sensation, they allow us to see in lower levels of illumination, *e.g.* moonlight, but the detail is poor. They also allow a general perception over an arc of about 190°.

It has long been believed that there are three different kinds of cone pigments which are responsive to the primary colours red, green and blue. The behaviour of the eye is consistent with this idea and relative responses for the three kinds of cones are shown in Fig. 1.17. The red cones show a maximum sensistivity at about 600 nm, the green cones around 550 nm and the blue at 460 nm. If a particular monochromatic yellow light denoted by A on the diagram is focused on to the retina it will stimulate the red and green cones with responses AC and AB. The red and green cones will relay the information to the brain where the sensation of yellow will be registered. It is thus logical to deduce that if red and green light simultaneously arrive at the retina, the stimulation of the red and green cones will give the impression of yellow which is not distinguishable from the monochromatic yellow. Thus yellow can be matched by a suitable mixture of the red and green primaries. It is, of course, the tricolour nature of colour vision that allows colour t.v. in its three-colour form to succeed. Integration of the three curves produces the overall response of the human eye shown in Fig. 1.7.

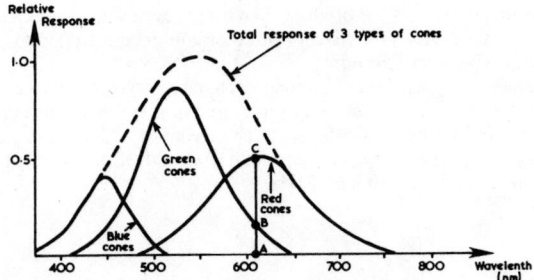

FIG. 1.17 APPROXIMATE RELATIVE RESPONSES OF THE THREE TYPES OF CONES

Colour Detail

If threads of different coloured cottons are held against a white background and viewed from about 10 metres it is difficult to name the actual colours. The threads themselves are quite discernible, *i.e.* the eye can perceive the fine detail but the colours show up as variations of light and dark.

Thus when colour appears in *fine detail* in a scene we are only conscious of the differing luminance levels and have difficulty in recognising colours. Because in a colour t.v. system valuable bandwidth is required to transmit the colour information, it would be wasteful to transmit the colour information of fine detail if it cannot be seen. This particular characteristic of human vision allows the colour information (hue and saturation) to be transmitted at narrow bandwidths without degrading the reproduced colour picture. On the other hand, the eye is very conscious of the luminance changes in fine colour detail, thus a comparatively wide bandwidth is required to transmit the luminance information.

Persistence of Vision

When the sensitive elements of the eye are stimulated by light from an object, the mental picture of that object does not disappear immediately if light from the object ceases to fall on the eye. The mental image tends to persist for a brief period of time (about $\frac{1}{25}$th second) and this effect is known as PERSISTENCE OF VISION. Use of this phenomenon is made in the projection of a continuous series of still shots in a cinema. The viewer does not observe the interruption between projected shots but sees a continuous picture. This is because the viewer retains a mental image of one shot and provided the following shot is projected before the image of the last one has disappeared, the viewer sees a continuous picture.

Persistence of vision is also put to use in the reproduction of television pictures on a receiver screen using a fast moving electron beam which traces out 25 pictures per second to give the effect of a continuous image. However, with this picture repetition rate there would be noticeable flicker present, particularly in high luminance parts of the picture. To overcome flicker a picture repetition rate of 50 per second is desirable. In television the problem of flicker is dealt with without resorting to 50 pictures per second by using a system of interlaced scanning, and this will be considered in Chapter 2.

Properties of light and human vision which have an important bearing on television may be summarised briefly as follows:

(a) Due to persistence of vision it is possible to give the effect of a continuous picture whilst interrupting the picture 25 times each second.

(b) The eye cannot distinguish between light that is coloured because it is of a particular wavelength and a mixture of lights of different wavelengths. Thus only three colours red, green and blue are necessary to produce most natural colours.

(c)   Certain colours appear brighter than others even when projected with equal energy. Any television system (mono or colour) must take this property of human vision into account.

(d)   The eye is very sensitive to luminance changes in a scene and a comparatively wide bandwidth (0—5·5 MHz) is required to preserve the fine detail. On the other hand the eye has difficulty in perceiving the colour of fine detail which this is a useful property as it means that less bandwidth is needed to handle the colour information (0—1·0 MHz).

CHAPTER 2

TELEVISION PRINCIPLES

THE essentials of a simple television system for closed circuit operation are shown in Fig. 2.1. The television camera lens focuses an image of the scene to be televised on the light sensitive surface of a camera tube. An electrical (video) signal voltage is obtained at the output of the camera with an amplitude proportional to the amount of light falling on the face of the camera tube. The output video signal is fed *via* a transmission cable to the display monitor where, after suitable processing and amplification, it is used to intensity-modulate the electron beam of the c.r.t. On the screen of the display device there is formed an optical image of the original scene. This

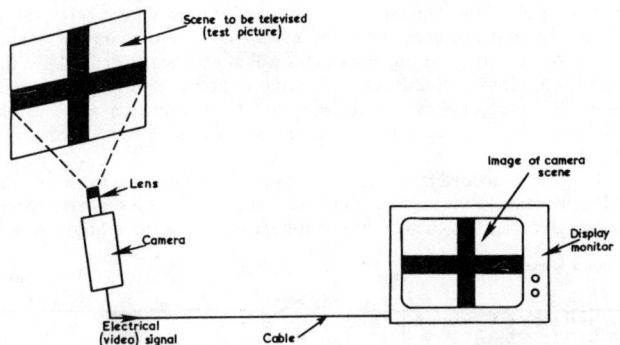

FIG. 2.1   THE BASIC ELEMENTS OF A TELEVISION SYSTEM (CLOSED CIRCUIT)

type of television system is called closed circuit t.v. (c.c.t.v.) because the video signal is confined to the cable, making the system 'private' as opposed to broadcast television which makes general reception possible.

It would appear from Fig. 2.1, and without a knowledge of camera operation, that the *entire* scene information is *simultaneously* conveyed to the monitor by the video signal. If this is correct, how can a *single* electrical voltage simultaneously represent the different light levels from various parts of the scene? Of course, it is not posssible. However, if there were *many* signal voltages produced simultaneously, with each generated voltage having an amplitude proportional to the amount of light emanating from a particular *small* area of the scene, the information could (in theory) be conveyed simultaneously. The breaking down of a scene into small areas or *elements* is an essential process in any television system and this will now be considered.

PICTURE ELEMENTS

Consider Fig. 2.2 (a) where the scene to be televised consists of a black cross on a white background, *i.e.* a monochrome scene. Imagine that the scene area is divided into 48

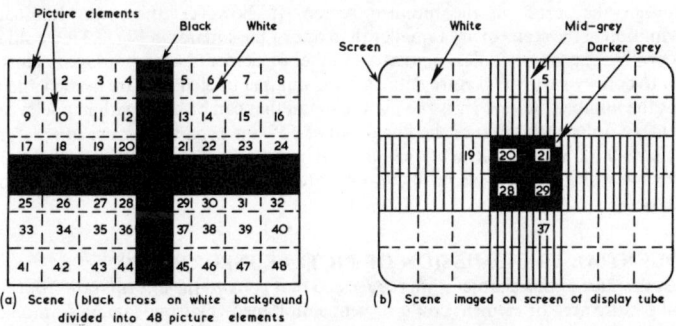

(a) Scene (black cross on white background)   (b) Scene imaged on screen of display tube
        divided into 48 picture elements

FIG. 2.2   BREAKING DOWN THE SCENE TO BE TELEVISED INTO PICTURE ELEMENTS
                (INSUFFICIENT NUMBER IN THIS CASE)

squares of equal size. With a monochrome scene there is only luminance information to be dealt with, thus each of the 48 squares contains a certain amount of luminance information. We will assume that the camera produces an output voltage of, say, 1 V for the white parts of the scene and 0 V for the black parts. Thus squares 1, 2, 15, 16, 38, 47 and 48, etc. which are completely white will each give rise to 1 V of output. Squares which are half white and half black (4, 5, 18 and 44, etc.) will each cause 0·5 V of output, and squares 20, 21, 28 and 29, which are only one quarter white, will each cause 0·25 V at the camera output. If the camera output voltage for each square is responsible for causing the brightening or darkening of a corresponding square on the screen of the picture monitor, the image of the black cross will not be clearly defined [see diagram (b)]. Squares 4, 5, 12, 13, 17 and 18, etc. will be reproduced as mid-greys and squares 20, 21, 28 and 29 as deeper greys, there being no sharp transition white to black as intended. This is because there are insufficient squares or elements to deal with the scene detail.

If each square is divided into four smaller squares or elements there will now be a total of 192 elements in the scene as illustrated in Fig. 2.3 (a). As before, assume that each element gives rise to a camera output voltage which is used to brighten or darken

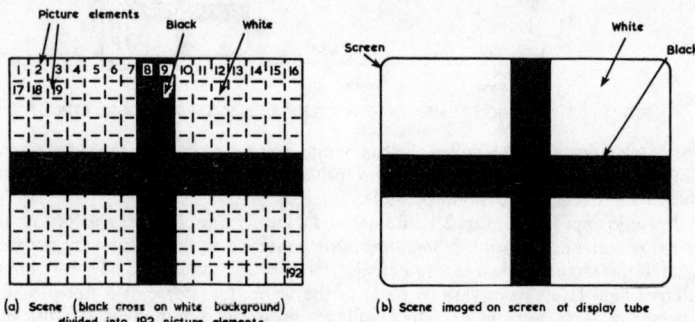

(a) Scene (black cross on white background) divided into 192 picture elements

(b) Scene imaged on screen of display tube

FIG. 2.3 INCREASING THE NUMBER OF PICTURE ELEMENTS (SUFFICIENT NUMBER IN THIS CASE)

corresponding elements on the monitor. There are now sufficient elements to deal with the scene detail, *i.e.* the picture on the monitor screen will show a sharp transistion from white to black on the vertical and horizontal edges of the black cross, see diagram (b). The improvement in detail occurs only if the edges of the image overlap picture elements as shown. The scene used in this example contains relatively large areas of constant luminance information, thus only 192 elements are required in our example to reproduce the scene on the monitor screen. If, however, there was luminance information in the scene of area smaller than one of the squares in Fig. 2.3 it would not be reproduced properly unless a greater number of elements were employed. Thus, in order to convey very fine picture detail a large number of elements are needed and the larger the number the smaller is the picture detail that can be dealt with in a television system. In order to reproduce the fine detail of 625-line broadcast television pictures, the number of picture elements required is approximately 5×10^5 elements. The manner in which the scene is broken down into picture elements will be explained in Chapter 3.

SEQUENTIAL TRANSMISSION OF PICTURE INFORMATION

At the start of this chapter it was suggested that *perhaps* the information from each small picture area or element could be sent simultaneously from camera to monitor. This clearly is not a practical proposition, as in 625-line broadcast television 5×10^5 separate lines or channels would be required between camera and monitor. Thus for technical and economic reasons a simultaneous system is not used. Instead a SEQUENTIAL method is used where the information contained in each picture element is 'read-off' in a logical sequence. At each 'reading' the camera tube produces an output VIDEO voltage having an amplitude proportional to the amount of luminance of each element (in a monochrome system). These sequential voltages are relayed from camera to monitor where the image is reconstituted element by element on the c.r.t.

The basic idea is shown in Fig. 2.4 where for simplicity only 20 picture elements are used. The picture elements of the camera image are read in sequence from left to right

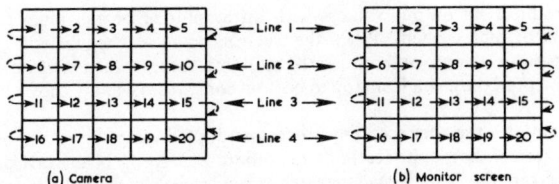

(a) Camera (b) Monitor screen

FIG. 2.4 SEQUENTIAL METHOD OF RELAYING PICTURE ELEMENT INFORMATION

starting at square 1 and proceeding to squares 2, 3, 4 and 5. At the end of the first row of squares (line 1), squares 6, 7, 8, 9 and 10 are 'read' completing line 2 and so on up to square 20. The process is then repeated commencing again at square 1. As each square is 'read' at the camera, the corresponding square on the monitor screen displays the relayed luminance information.

Clearly, the elements must be 'read' over and over again in RAPID SEQUENCE in order to create a continuous picture on the monitor screen and to capture any movement within the scene. As long as complete scene 'readings' are taken often enough, say 25 per second, PERSISTENCE OF VISION will ensure a continuous picture on the monitor screen.

Should camera and monitor be out of step in 'reading' and displaying information the reproduced picture may be a complete jumble. For example, suppose that the information of camera square 1 is displayed in the space allocated to square or element 11 on the monitor. Due to the sequential manner of relaying and displaying the information, the information of camera squares 1—5 will be displayed in monitor squares 11—15, and the information of camera squares 6—10 will be displayed in monitor squares 16—20. Camera information from squares 11—20 will thus be displayed in monitor squares 1—10, *i.e.* the bottom half of the camera scene image now appears at the top of the monitor screen and *vice-versa*. Clearly, to keep camera and monitor in step there must some form of SYNCHRONISM between them.

SCANNING

In practice the 'reading' and displaying of element information is carried out by scanning electron beams inside the camera and display tubes. The idea of an elementary scanning system is shown in Fig. 2.5, only eight scans being shown for simplicity. Inside the camera tube an electron beam traces out a series of sloping lines

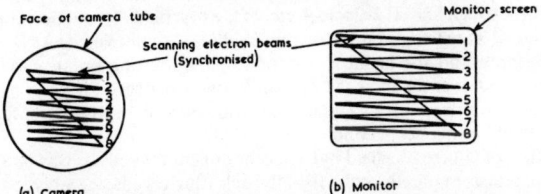

(a) Camera (b) Monitor

FIG. 2.5 USE OF ELECTRON BEAM FOR SEQUENTIAL 'READING' AND DISPLAY OF ELEMENT INFORMATION (SIMPLE SCANNING)

across the face of the camera tube. As the beam is moving across the tube a force is exerted on the beam causing a downward movement (this is responsible for the downward slope of the lines).

The beam commences scanning at the left-hand side of the camera face and traces out line 1. At the end of line 1, the beam rapidly returns to the left-hand side to commence the next LINE SCAN (line 2). At the end of line 2 the beam again rapidly returns to the left-hand side to commence line 3 and so on. Upon completing line 8, the full image formed on the face of the camera tube by the lens system has been scanned. In carrying out this process the electron beam examines each element in turn and the camera produces an output voltage with an amplitude proportional to the light information of each element. When the electron beam reaches the end of line 8, it quickly returns to the top and repeats the scanning process starting again with line 1. The movement of the beam from top to bottom constitutes what is known as the FIELD SCAN.

At the monitor the same scanning process is adopted using an electron beam which scans the screen of the monitor cathode-ray tube (c.r.t.). As the beam travels across and down the screen, the beam is made greater or less intense by the voltage output of the camera causing a larger or smaller light output from the screen.

Interlaced Scanning

The scanning beams in the camera and monitor must repeat the scanning process many times per second to give the impression of a continuous picture to the viewer. Persistence of vision lasts for approximately $\frac{1}{25}$th second. If a picture repetition rate of 25 pictures per second were used the eye would perceive an uninterrupted picture on the screen. There would, however, still be some flicker particularly in the bright areas of the picture. To overcome flicker it is necessary to increase the picture rate to 50 pictures per second. This speeds up the whole scanning process, *i.e.* more information has to be packed into a shorter time interval, resulting in a doubling of the video bandwidth required. The transmission bandwidth available to the Television Broadcast Authorities has always been at a premium. In the early days of monochrome television transmissions alternative ways were investigated to get over the flicker problem without resorting to 50 pictures per second. This led to the introduction of INTERLACED SCANNING, a system which is used internationally for broadcast television and in the majority of CCTV systems.

In an interlaced system an odd number of lines must be used, *e.g.* 625 lines, 525 lines, etc. An example of an 11-line system is shown in Fig. 2.6.† The complete picture is divided into two distinct FIELDS with each field containing $5\frac{1}{2}$ lines. Scanning commences at *A* with the beam tracing out lines 1, 2, 3 and 4, etc. At the end of each line, LINE FLYBACK occurs causing the beam to rapidly return to the left-hand side of the camera and monitor tubes. Due to the effects of the FIELD SCAN, the beam is also being deflected downwards as previously explained. Half-way through line 6 the first (odd) field is completed and FIELD FLYBACK commences at point *B*. The beam now returns to the top of the screen and the camera tube. Assuming that the field flyback is instantaneous, the other half of line 6 will now be completed as shown when the second (even) field commences at point *C*. Because of the half-line, scanning lines 7, 8 and 9, etc. now fit in between lines 1, 2, 3 and 4, etc. of the first field. This action continues until the end of line 11 at which point the second field scan is completed. At *D*, field flyback occurs once more and the beam is returned to the starting point *A*. This action is repeated over and over again at the rate of 25 times per second. Each picture is thus composed of 2 interlaced fields (interlaced field: picture ratio of 2:1) and therefore there are 50 field scans per second.

In practice, of course, the field flyback is not instantaneous as was assumed in Fig. 2.6.† A finite time is required for the field flyback which is considerably longer than the line flyback period. During field flyback the circuits causing line deflection, being continuous in operation, still cause the beam to take a horizontal zig-zag path across the screen and camera tube. Fig. 2.7† shows the path taken by the beam during field flyback, but the beam is not normally visible during these periods.

If the monitor (or receiver) brightness control is turned up when there is no picture modulation, the scanning lines produced by the c.r.t. electron beam can be seen. The pattern of lines visible on the screen is called the RASTER.

Interlaced scanning tricks the eye into believing that every part of the scene is repeated 50 times per second when in fact each element of the scene is only being flashed up on the screen 25 times per second. The difference between interlaced and non-interlaced scanning, and the effect on human vision, may be seen by considering Fig. 2.8 which shows a small area of picture detail (about two beam widths from top to bottom). With interlaced scanning, line x on, say, the odd field causes the top half of the

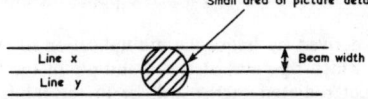

FIG. 2.8 OPTICAL 'TRICK' OF INTERLACED SCANNING

detail to brighten-up, and $\frac{1}{50}$th of a second later on the even field, line y causes the lower half of the detail to brighten-up. Although it takes $\frac{1}{25}$th of a second for the whole of the detail to be shown, a part is being flashed up on the screen every $\frac{1}{50}$th of a second. In a non-interlaced, 25 pictures per second system, lines x and y will occur in the same field. Thus the upper and lower portions of the picture detail will brighten up in the time interval of two consecutive lines, but will not brighten up again until $\frac{1}{25}$th of a second later on the following field scan.

BASIC TRANSMITTER AND RECEIVER ARRANGEMENTS

Basic block diagrams of the vision transmitter and television receiver for monochrome operation are given in Figs. 2.9 and 2.10, now to be described. In the camera (A), light from the scene is focused by the lens on to the face of the camera tube

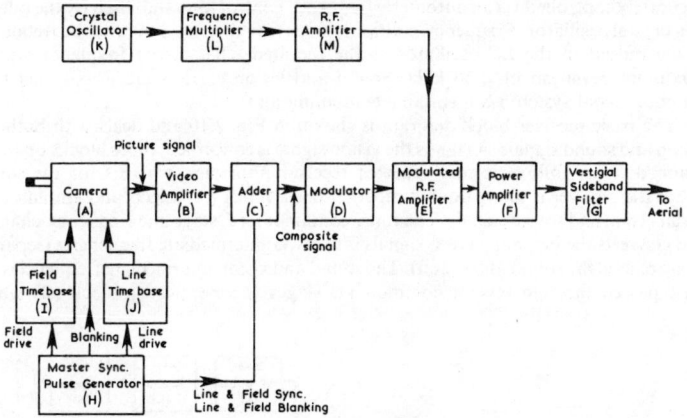

FIG. 2.9 BLOCK DIAGRAM OF VISION TRANSMITTER (MONOCHROME)

where an *electrical charge image* of the scene is formed. The camera tube electron beam is deflected over the photosensitive face of the tube by sawtooth currents supplied to the camera deflector coils from the LINE TIMEBASE (J) which produces the horizontal scan and the FIELD TIMEBASE (I) which produces the vertical scan. The timing of the camera's timebases is controlled by the master sync. pulse generator (H). This unit serves as the master timing for the complete scanning process at transmitter and receiver. To ensure that camera and receiver display tube scans are correctly synchronised, line and field drive pulses are fed to the camera timebases, and line and field sync. pulses are fed to the adder (C) for subsequent use in synchronising the receiver timebases. In addition, block (H) supplies BLANKING waveforms to remove any spurious signal generated during the line and field flyback periods in the camera, also blanking to be added to the composite waveform in block (C).

As the electron beam scans the charge image in the camera. it 'reads' each element in turn producing an electrical signal voltage (video signal) at the camera output. Since the output voltage of the camera is small it is amplified in block (B) and then fed to block (C) where sync. and blanking are added to the video signal. Thus, at the output of block (C) we have the *composite* signal consisting of video, sync. and blanking. The output of the adder is of the order of 1 V and this may have to be raised to 100 V or so in order to AMPLITUDE MODULATE the r.f. carrier in block (E). The amplification to this required level is carried out in block (D) which supplies the modulating signal to the modulated r.f. stage (E).

The r.f. carrier is derived by frequency multiplication of the output of a stable crystal oscillator (K) which operates at a submultiple of the final radiated carrier frequency. As the final radiated carrier will lie in the u.h.f. band the frequency multiplier block (L) will contain several multiplier stages since the crystal oscillator operates at low frequency and frequency multiplication is usually limited to × 2 or × 3 per stage.

In block (E) the composite video signal from the modulator causes amplitude modulation of the amplified r.f. carrier from block (M). The modulated carrier is then fed to block (F) which supplies the necessary r.f. power for delivery to the aerial system. Before reaching the aerial the vision transmission is passed through a filter which removes part of the lower sideband in order to conserve bandwidth. This is called a 'Vestigial sideband filter' since it leaves only a 'trace' or 'vestige' of the lower sideband (vestigial sideband transmission is dealt with more fully in Chapter 5.).

The sound section of the transmitter follows conventional lines for FREQUENCY MODULATION. The actual modulation is carried out on a low frequency carrier by modulating a low frequency *LC* oscillator. The centre frequency of the *LC* oscillator is accurately controlled *via* an automatic frequency control loop stabilised by the output of a crystal oscillator. Frequency multiplier stages are then used to raise the frequency of the output of the *LC* oscillator to the required u.h.f. carrier frequency with a maximum deviation of ±50 kHz. Sound and vision carriers are usually fed to a common aerial system *via* a suitable combining unit.

The basic receiver block diagram is shown in Fig. 2.10 and deals with both the vision and sound signals. As far as the vision signal is concerned, those blocks up to the vision detector follow normal superhet receiver principles, whereas for the sound signal the receiver is along double superhet lines. Block (A) selects and amplifies the aerial vision and sound signals. The tuner contains an r.f. stage and frequency-changer and converts the incoming u.h.f. signals into lower intermediate frequencies (separate i.f.s. are used for sound and vision). The sound and vision intermediate frequencies are then passed through several common i.f. stages represented by block (B) which

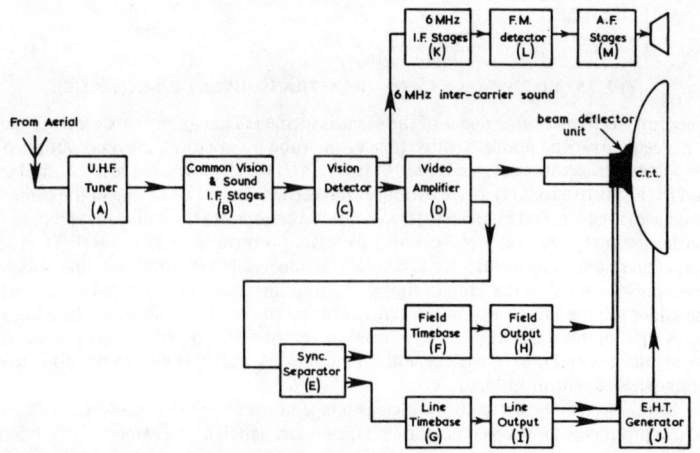

FIG. 2.10   BASIC BLOCK DIAGRAM OF 625-LINE TELEVISION RECEIVER (MONOCHROME)

provides most of the gain and selectivity of the receiver. In block (C) the vision signal is detected, filtered and then passed on to the video amplifier. Also in the vision detector (although a separate detector is sometimes used) the sound i.f. is converted into a lower i.f. of 6 MHz called the 'intercarrier sound i.f.' The intercarrier i.f. is produced by the 'beating' together of the normal vision and sound i.f.s in the vision detector, with the detector acting as a mixer and the vision i.f. as a kind of local oscillator signal. The 6 MHz beat which contains the frequency modulation of the original sound signal is then amplified in block (K), demodulated in block (L) and the resulting audio signal amplified in block (M). The reasons for using the intercarrier system will be dealt with more fully later.

Returning to the detected vision signal which is applied to the video amplifier (D), this stage amplifies the video signal to a level sufficient to vary the intensity of the scanning electron beam inside the c.r.t. The output of the video amplifier also contains the vital line and field synchronising pulses which are separated from the picture signal in the sync. separator (E). The sync. pulses are now separated into line and field sync. pulses and fed to the corresponding timebases (G) and (F). The timebase outputs are then applied to output stages which supply the necessary scanning waveforms to the beam deflector unit on the c.r.t. The c.r.t. also requires an e.h.t. supply which is provided by block (J) fed from the line output stage (I). In addition to the features shown in Fig. 2.10 a t.v. receiver will use some form of a.g.c. (and often automatic frequency control of the local oscillator in the tuner) and will require a power supply.

DEFLECTING THE ELECTRON BEAM

In both camera and receiver the electron beam is deflected in two directions at right angles to each other. This is done by magnetic fields set up at right angles to one another using two pairs of scan coils (LINE and FIELD SCAN COILS). The scan coils are mounted around the necks of the display c.r.t. and camera tube so that their resulting magnetic fields pass through the glass necks to influence the path taken by the electron beam. Fig. 2.11(a) shows the disposition of the magnetic fields produced by the line and field coils in a display c.r.t. The horizontal magnetic field pattern produced by the field

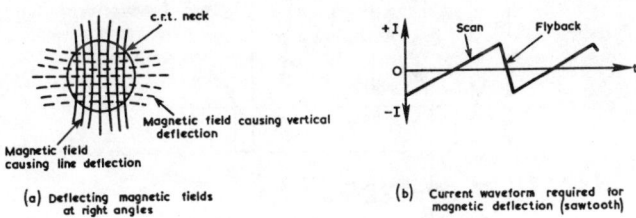

(a) Deflecting magnetic fields at right angles

(b) Current waveform required for magnetic deflection (sawtooth)

FIG. 2.11   MAGNETIC DEFLECTION

scan coils causes vertical deflection of the beam, and the vertical magnetic field pattern of the line scan coils causes horizontal deflection of the beam. In modern camera tubes the movement of the electron beam in relation to the disposition of the scan coils is not the same as for the receiver c.r.t. due to the presence of another magnetic field set up by a long focus coil (the details of which need not concern us here) but follow similar basic principles.

In either the horizontal or vertical directions the movement of the beam must be at a LINEAR RATE during the SCAN and RAPID later on during the RETURN OR FLYBACK. This means that the magnetic field must be increased gradually in a linear fashion during the scan and then rapidly reduce to its 'start of scan' value during flyback. To achieve this, currents having sawtooth waveshapes have to be fed into the scan coils, Fig. 2.11(b).

Scanning current waveforms for interlaced scanning using an 11-line system (for simplicity) are illustrated in Fig. 2.12. Here it has been assumed that the field flyback

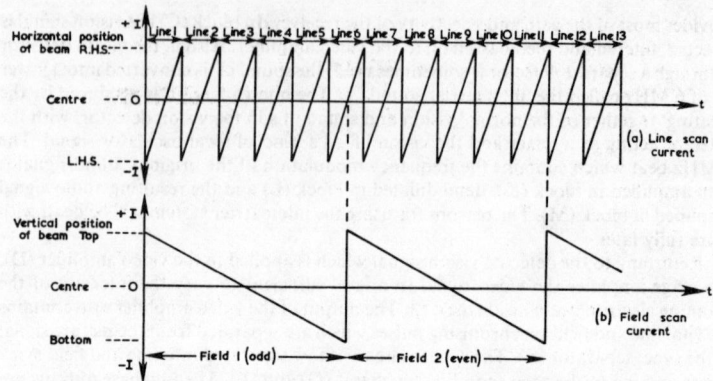

FIG. 2.12   SCANNING WAVEFORMS FOR AN 11-LINE INTERLACED SYSTEM

(and line flyback) is instantaneous. In practice, the field flyback time is about $\frac{1}{20}$th of the field scan time. Within limits the duration of the field flyback is quite arbitrary and does not have to be an exact number of line durations as at first may be thought. Provided the field flyback time is the same on odd and even fields, satisfactory interlacing of the scanning lines will occur.

THE VIDEO SIGNAL AND SYNCHRONISING PULSES
Voltage Range

The composite video signal for the British 625-line system over a period of one line is shown in Fig. 2.13. The waveform is divided on an amplitude basis into two voltage ranges. The range between 28% and 100% is reserved for the picture signal and the

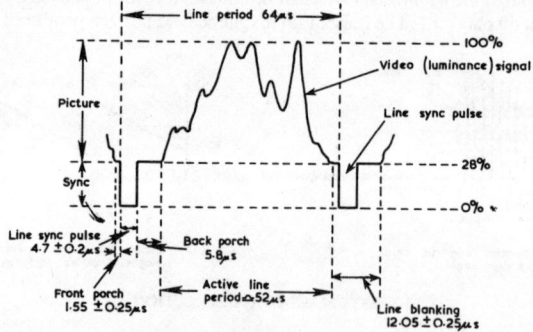

FIG. 2.13   LUMINANCE SIGNAL WAVEFORM FOR 625-LINE SYSTEM (LINE DIMENSIONS)

range between 0% and 28% of the complete waveform is taken up by the sync. pulses. For broadcast television the ration of picture/sync. is about 2·5:1. This ratio is chosen so that when at some remote receiver where the signal level is weak, the timebases will fail to synchronise at the same time as the picture signal-to-noise ratio falls to such a level that the picture ceases to be of viewable quality. In monochrome t.v. system 28% corresponds to BLACK LEVEL and 100% to PEAK WHITE. For colour transmissions, these percentage levels should be considered as representing zero and maximum luminance respectively.

The video signal can have any arbitrary value between 28% and 100% and on a normal scene each line of video information will be different fron the preceding one. When 'scoping' the line waveform of such a scene using an ordinary c.r.o., the display will show SUPERIMPOSED lines of information. As the video content is different line-to-line, the displayed video will appear very fuzzy. The line sync. pulses, being of repetitive shape will be quite distinct. When the scene information is repetitive line-by-line as with some test pictures, *e.g.* a 'grey-scale', the video information of the displayed waveform will also be distinct.

Line Dimensions

With 625 lines per picture and 25 pictures per second, the *line frequency*
$$= 625 \times 25 = 15,625 \text{ Hz}$$

$\therefore$ the line period $= \dfrac{1}{15625}\text{s} = 64\mu\text{s}.$

Not all of the 64μs line period is active, *i.e.* carrying picture information. During what is termed the 'line blanking period' no picture information is transmitted. Any spurious signals generated by the camera between the end of one scanning line and the commencement of the next line are suppressed by means of a 'blanking pulse' of suitable amplitude and duration which is added to the video signal. By using a limiter to limit the pulse at blanking level, the spurious signals are lost in the limiter. The blanking pulse does not appear in the composite waveform, but this non-picture information period is given the name 'blanking period' because of the blanking pulse used to create it. The line blanking period can be divided into three sections.

(a) Front Porch

This is the brief period (1·5μs) just before the commencement of each line sync. pulse. It serves as a 'cushioning' period for the video circuits in the camera and receiver, allowing them to settle down before the commencement of the regularly occurring sync. pulses. The level of the video signal at the end of a line of picture information is quite arbitrary. Electronic circuits cannot change their voltage state instantaneously and the front porch allows sufficient time for the voltage level to fall from 100% to the 28% level before the line sync. pulse commences. Without this interval, the sync. pulses may be late in starting following lines ending at high voltage level compared with the pulses following lines ending at low voltage level, resulting in faulty line timebase synchronisation.

(b) Line Sync. Pulses

This pulse provides the timing of the line timebase oscillations in the receiver, thus synchronising them with similar oscillations in the camera. The *leading* and downward-going edge of the pulse initiates the commencement of the line timebase *flyback*. A duration of 4·7μs is allowed for the pulse and during this interval the electron beams in camera and receiver will be in a retrace stroke.

(c) Back Porch

The back porch provides a further 'grace' period for the line timebase in the receiver to complete its flyback before picture information recommences on the next line. This is necessary due to variations in design of the line output stage used in different makes of receiver. The back porch may also be used as a brief sampling period for measuring the amplitude of the sync. pulses for a.g.c. purposes in the receiver.

Field Sync. Pulses

With 25 pictures per second and 2 fields per picture, the FIELD FREQUENCY

$$= 2 \times 25 \text{ Hz} = 50 \text{ Hz}$$

$\therefore$ the FIELD PERIOD $= \dfrac{1}{50}\text{s} = 20$ ms.

Field synchronising pulses at a rate of 50 per second are thus required to synchronise the field timebase in the receiver. It might at first be thought that a single long pulse of,

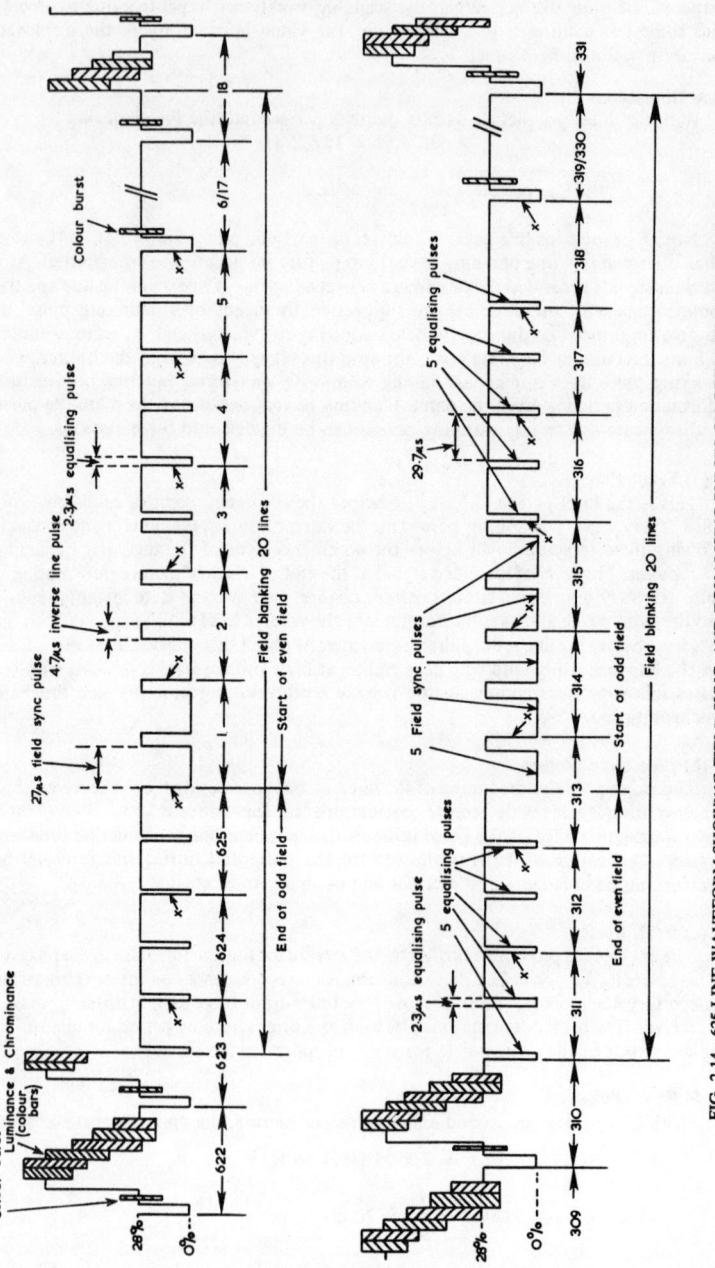

FIG. 2.14   625-LINE WAVEFORM SHOWING FIELD SYNC. AND EQUALISING PULSES ON ODD AND EVEN FIELDS

say, 1 ms could be used. Although this will work and is used in industrial CCTV systems it is not entirely satisfactory since during the period of the pulse there are no sync. pulses to keep the line timebase in synchronism. This may appear to be unimportant, but without any line sync. during the field flyback period the line timebase may drift in frequency causing the first few lines at the top of the picture to be laterally displaced. Thus a system must be used which maintains the line timebase in synchronism during the field flyback period. A further complication arises due to the use of interlaced scanning, the conditions being dissimilar at the end of alternate fields. Fig. 2.14 shows the complex 625-line field waveform which overcomes the difficulties mentioned above. The essential features are as follows.

(a) Field Sync. Pulses
At the end of each field there are five broad 27 μs pulses occurring at half line intervals. In the receiver these pulses are used to build up a field locking pulse which initiates the flyback stroke of the field timebase oscillator. It should be noted that the fields are labelled differently to those shown in Fig. 2.12. The start of the EVEN field begins on line 1 and the start of the ODD field half-way through line 313. The ODD field is by definition the one in which picture information ends half-way through a line (line 623). Although this appears confusing, it does not affect the principle of interlaced scanning and one should merely note the labelling of the lines and fields that are adopted.

(b) Equalising Pulses
A group of five 2·3 μs equalising pulses are added before and after the field sync. pulses in each field. These pulses are included to ensure that the field locking pulse generated in the receiver has precisely the same shape on odd and even fields. The need for these pulses is because of the dissimilar conditions at the end of odd and even fields.

(c) Field Blanking Period
In each field, no picture information is transmitted during the two groups of equalising pulses (5 lines in total), the field sync. pulses (2½ lines) and for a further 12½ lines. Thus, picture information is suppressed for 20 lines each field. This period is called the 'field blanking period' and allows sufficient time for the field timebase in the receiver to complete its flyback stroke. In each COMPLETE PICTURE PERIOD there are therefore 40 lines not bearing picture information, i.e. the total number of ACTIVE LINES is 625 − 40 = 585 lines. During the field blanking period, lines 19 and 20 on one field and lines 332 and 333 on the next field are used for sending out test signals. In addition, lines 17 and 18 and lines 330 and 331 are used for transmitting a series of coded electronic pulses for data transmissions (CEEFAX and ORACLE).

(d) Line Timebase Synchronism
It will be noted from Fig. 2.14 that at the end of each line period there is a downward-going edge of a pulse (marked x). These pulse edges, after suitable processing in the receiver, are used to maintain the receiver line timebase in full synchronism during the field blanking interval.
In order to observe the field sync. pulses on an ordinary c.r.o., it should be synchronised at picture frequency (25 Hz) and should be capable of large X expansion. Detailed examination of the field sync. pulse train is best carried out using a c.r.o. with a strobe timebase facility (if available).
At one time the field frequency was locked to the mains supply frequency (synchronous working) but not any longer, therefore it can be maintained accurately at 50 Hz (50 ± 0·05 Hz).
In Fig. 2.14, the chrominance (colour) information of the non-suppressed lines has been included as well as the colour burst. The significance of these signals will be shown at a later stage.

PICTURE RESOLUTION

The 'resolution' or 'definition' of a t.v. system is a measure of its ability to deal with fine picture detail, *i.e.* to reproduce a sharply-defined image of small detail in the scene. Normally, we are interested only in the maximum resolution of a system which gives an indication of the finest detail that a system can handle.

The definition of any reproduced picture depends upon both the HORIZONTAL and VERTICAL RESOLUTION. Horizontal resolution is a measure of the ability to reproduce luminance changes along a horizontal line such as *xy* in Fig. 2.15(a). In this diagram vertical black and white stripes are assumed to be imaged on the face of the camera

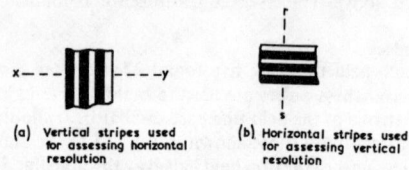

(a) Vertical stripes used
for assessing horizontal
resolution

(b) Horizontal stripes used
for assessing vertical
resolution

FIG. 2.15   HORIZONTAL AND VERTICAL RESOLUTION

tube. If the system has insufficient horizontal resolution, the luminance changes along any horizontal line scan will appear blurred, *i.e.* the sharp changes from black to white will not be reproduced. The finer the vertical stripes, the faster the system will have to respond to reproduce them clearly. In any reproduced picture, HORIZONTAL RESOLUTION shows up on the VERTICAL EDGES of picture detail and depends upon the frequency bandwidth of the system. Vertical resolution is a measure of the ability to reproduce fine detail along a vertical line such as *rs* in Fig. 2.15(b). Here, horizontal black and white stripes are being considered. To reproduce fine detail in a vertical sense, a large number of scanning lines are required and the more there are the better the vertical resolution. In any reproduced picture, VERTICAL RESOLUTION shows up on the HORIZONTAL EDGES of picture detail.

Vertical Resolution

In a 625-line system there are 585 active lines as we have seen, and the vertical resolution would appear to be that obtainable from this number of lines, but in practice it is less that 585 lines. It is important to note that the relative positioning of the scanning beam in the camera to horizontally disposed picture detail affects the resolution. Consider Fig. 2.16(a) which shows an electron beam scanning a horizontal stripe pattern where the thickness of each stripe is equal to the diameter of the electron beam. If the centre of the electron beam aligns with the centre of each stripe, the pattern can be clearly resolved. However, if the centre of the beam aligns with an edge of a stripe as in diagram (b) the pattern cannot be resolved as the beam will 'read' the average of the stripes which is grey. With the beam in intermediate positions between

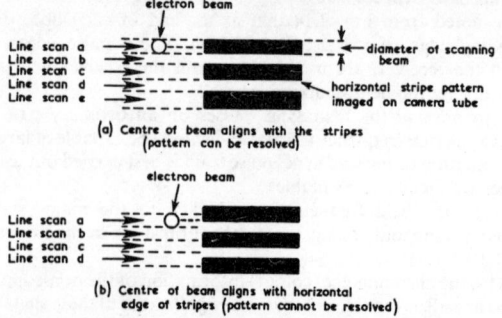

(a) Centre of beam aligns with the stripes
(pattern can be resolved)

(b) Centre of beam aligns with horizontal
edge of stripes (pattern cannot be resolved)

FIG. 2.16   EFFECT OF BEAM ALIGNMENT WITH ULTIMATE STRIPE PATTERN

these two extremes there will be some loss of resolution and the patt(rn will not be so clearly defined.

The position of the scanning beam in relation to such a test pattern (or similar picture detail) will depend upon how accurately optical positioning and scan geor.·.:try is set and maintained. Statistical and subjective testing suggests that the effective number of lines is about 0·7 of the total number of active lines. This factor is known as the 'Kell factor' (after an early television worker). The Kell factor cannot be precisely measured so values other than 0·7 may be used. Assuming a Kell factor of 0·7, the maximum vertical resolution obtainable in a 625-line system is 0·7 × 585 = 409·5 lines (say 410 lines).

Horizontal resolution

An obvious aim in designing a t.v. system would appear to be to make the horizontal resolution the same as the vertical resolution. With a maximum vertical resolution of 409·5 lines per active picture height, the horizontal resolution required to match it is also 409·5 lines per active picture height.

Now the horizontal resolution is determined by the video bandwidth, and to find out what bandwidth is needed consider Fig. 2.17. This diagram shows a checker-board pattern consisting of alternate black and white squares which is to be reproduced on

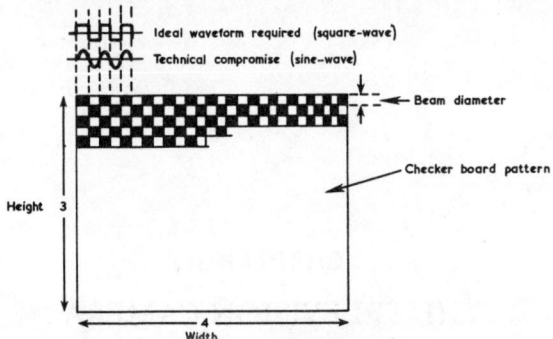

FIG. 2.17 CHECKER-BOARD PATTERN TO DETERMINE HIGHEST VIDEO FREQUENCY REQUIRED

the screen of the t.v. receiver. There are 409·5 squares along the vertical side of the board, this number representing the maximum practical vertical resolution. Each square has sides corresponding to the diameter of the electron beam and represents the ultimate pattern for testing the resolution.

In television the ratio of Picture Width : Picture Height (called the 'aspect ratio') is 4:3. Therefore the number of squares across the board in a single horizontal line is 4/3 × 409·5 = 546. At each line scan this number of squares has to be scanned by the beam in a time period of 52 μs (the active line period). To reproduce such a pattern, the video waveshape required is ideally a square wave. To handle this waveshape would place too great a demand on the system in terms of bandwidth. A reasonable technical compromise is to accept a sine wave response. Note, however, that each cycle corresponds to TWO squares. Thus the number of cyclic changes across each picture line is 546/2 = 273.

Therefore the periodic time of one cyclic change $= \dfrac{52}{273} \mu s = 0\cdot19 \ \mu s$

Thus the frequency of the cyclic changes $= \dfrac{1}{0\cdot19 \times 10^{-6}} \ Hz. = 5\cdot25 \ \textbf{MHz.}$

This figure is close to the upper video frequency of 5·5 MHz transmitted in the 625-line signal. Thus the following expression may be used for determining the highest video frequency:

$$\text{Highest video frequency} = \frac{\text{no. of active lines} \times \text{aspect ratio} \times \text{Kell factor}}{2 \times \text{duration of one active line period}}$$

CHAPTER 3

THE TELEVISION CAMERA AND OUTPUT SIGNALS

A t.v. camera usually consists of:
  (a) A vacuum camera tube.   (c) Scanning generators.
  (b) An optical system of lenses. (d) Video amplifiers.

The camera tube develops picture signals by scanning a 'charge image' derived from an 'optical image' (*via* the lens system) of the scene to be televised. The camera tube has two essential functions: (a) it must subdivide the optical image into a large number of small 'elements'; and (b) it must scan these elements in a certain order and at each one produce an electrical (video) output signal with an amplitude corresponding to the amount of light falling on that element.

For many years camera tubes using two different operating principles have been available. In one type, the electrode on which the optical image is formed is photoemissive, *i.e.* coated with a material which liberates electrons on exposure to light. These tubes have been used extensively in high-definition studio work. Other tubes use a photoconductive material, the electrical resistance of which is lowered on exposure to light. The photoconductive principle is used in the vidicon and plumbicon camera tubes which will be described.

THE MONOCHROME CAMERA

The general layout of a vidicon camera tube which may be used in a monochrome t.v. camera is shown in Fig. 3.1. This is available in different sizes but a common type is

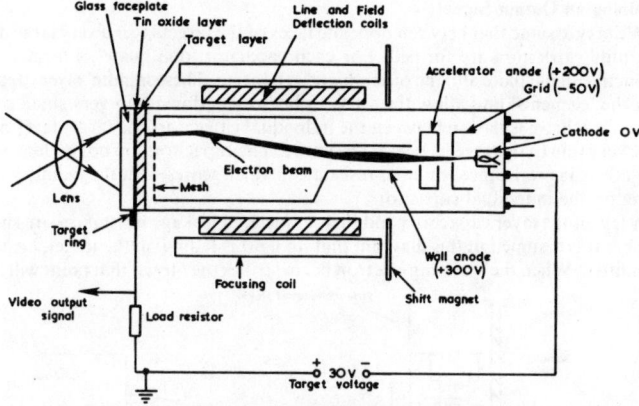

FIG. 3.1   THE VIDICON CAMERA TUBE

about 6 inches long with a diameter of about 1 inch. At one end is a photoconductive target and an electron gun assembly at the other.

The electron gun shoots a narrow beam of electrons along a highly evacuated glass envelope. Electrons are accelerated towards the target by the accelerating anode, held some 200 V positive with respect to the cathode. Before reaching the accelerator anode, the electrons must pass through the grid disc. This electrode is held negative with respect to the cathode and its bias voltage controls the beam intensity. After passing through the accelerating anode, the electron beam must be brought to a sharp focus, either by varying the potential of the wall anode or by varying the current in the focus coil. To provide horizontal and vertical deflection of the beam so that it scans the target image, sawtooth currents are fed into the line and field deflector coils.

Over the face of the wall anode is a fine mesh through which most of the arriving electrons pass. The electric field set up between the mesh and the target layer provides a uniform decelerating field for the electrons so that they strike the target at low velocity. Under the influence of the *combined* deflection and focusing fields the electron beam is deflected laterally (up or down) but on coming under the influence of the focusing field only, the beam travels in a straight line path so that it strikes the target at right angles wherever it lands. This is known as 'orthogonal scanning' and, combined with the decelerating field, ensures minimum secondary emission from the target layer which reduces the generation of spurious signals from the camera.

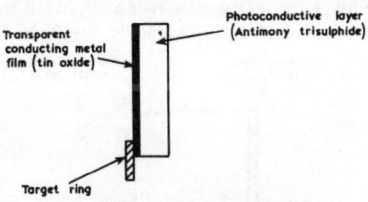

FIG. 3.2   TARGET DETAILS

Close to the mesh is the target, consisting of a layer of photoconductive material such as antimony trisulphide, whose resistance decreases when light falls on it. One face of the target layer has a thin transparent film of a conducting medium such as tin oxide which is electrically connected to a metal ring sealed in the glass envelope. The 'target ring' forms the signal output electrode. Light from the scene is focused by the lens system through an optically flat polished glass faceplate and the tin oxide layer on to the target material.

Obtaining an Output Signal

We may assume that between opposite faces of the target layer a very large number of minute capacitors are formed. For each capacitor, one 'plate' is formed by the conductive signal plate and the other 'plate' is floating. These minute 'layer capacitors' form the 'elements' and allow the scene to be broken down into very small separate areas. To achieve isolation between the individual capacitors the target layer must be made very thin to increase its lateral resistance. The capacitors are not perfect, *i.e.* they have a leakage resistance (or axial resistance) which varies with the quantity of light falling on the individual capacitors.

A few of the layer capacitors and their associated leakage resistances are shown in Fig. 3.3. It is assumed in this diagram that no light is falling on the target, *i.e.* the lens cap is fitted. When the scanning electron beam strikes the target, that point will assume

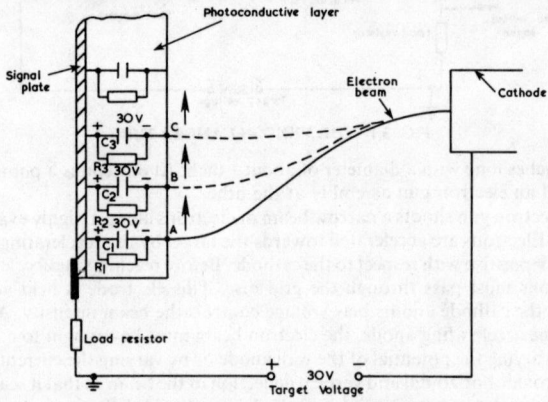

FIG. 3.3 STATE OF CHARGE OF A FEW 'LAYER CAPACITORS' AFTER FIRST SCAN BY ELECTRON BEAM (NO LIGHT ON TARGET LAYER)

cathode potential (the beam acting like a conductor) and each layer capacitor will charge to the target voltage on the first scan. Between scans each capacitor will discharge a little *via* its leakage resistance. However, the charge lost will be replaced on the next and subsequent scans when the capacitors recharge in turn. While recharging, a minute current will flow in the load resistor which is known as the 'dark current' of the tube (typically 0.01 μA).

If light is allowed to fall on the target as shown in Fig. 3.4, high intensity light falling on point A' causes the resistance R_1 to fall, causing C_1 to discharge to, say, 25 V. Low intensity light falling on point B' causes the resistance of R_2 to fall but the reduction in resistance is less than for R_1. As a result C_2 discharges to, say, 29 V. As no light falls on

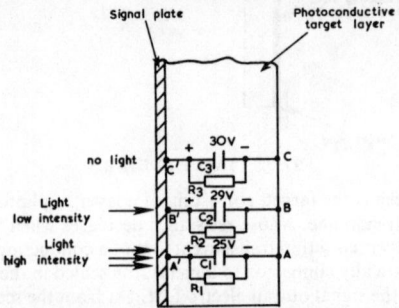

FIG. 3.4 LIGHT FALLS ON TARGET LOWERING RESISTANCES R_1 AND R_2, CAUSING C_1 AND C_2 TO DISCHARGE

point C' the layer capacitor C_3 does not discharge. Thus it can be seen that the charge lost by each capacitor is proportional to the quantity of light falling on those points, and the scene image has been converted into a 'charge image'.

When the electron beam subsequently scans the charge image each capacitor in turn is recharged (if necessary) back to 30 V. Fig. 3.5 shows the beam striking point A causing C_1 to recharge from 25 V to 30 V. As the capacitor charges, current flows in the

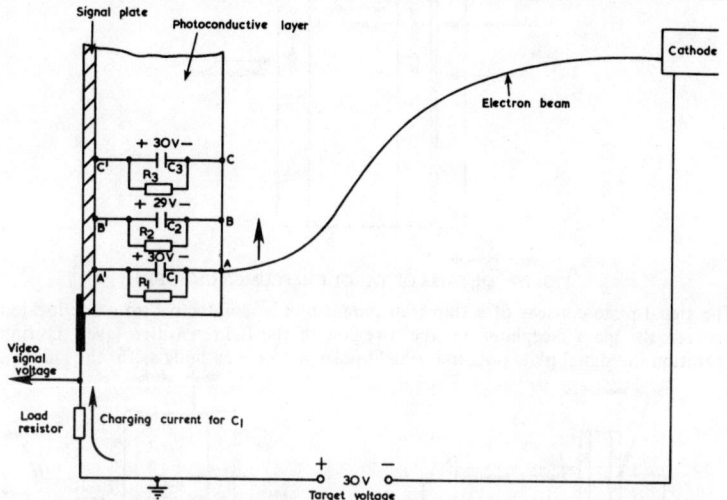

FIG. 3.5 SUBSEQUENT SCANS CAUSE C_1 AND C_2 TO CHARGE RESULTING IN A CHARGING CURRENT FLOW IN THE LOAD RESISTOR

load resistor and the p.d. developed across the resistor constitutes the output signal. As the beam scans over the entire target area each layer capacitor is recharged in sequence and each time the charging current, which is proportional to the intensity of the light falling on the point being scanned, causes an output signal (video) voltage across the load resistor. The peak current in the load resistor is typically $0.3\ \mu A$. Fig. 3.6 shows the type of signal voltage to be expected across the load resistor.

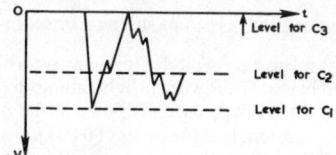

FIG. 3.6 VIDEO SIGNAL VOLTAGE ACROSS LOAD RESISTOR

One of the main disadvantages of the vidicon tube is that it takes time for the target material to change its resistance when varying intensity light falls on the target layer (referred to as the 'vidicon lag'). This causes some smearing of reproduced moving images particularly when the light level is low. For this reason, the vidicon is best suited where there is little rapid movement in the scene.

A more modern photoconductive tube is the plumbicon which has a low-lag characteristic. The plumbicon, like the vidicon, uses a low velocity electron beam and orthogonal scanning. The main difference lies in the make-up and operation of the target layer; see Fig. 3.7.

In the plumbicon tube the layer on which the scene is imaged consists of microcrystalline lead monoxide, hence its name. The lead monoxide layer is made up of a thin p-type region, a thick intrinsic lead monoxide region and a thin n-type region.

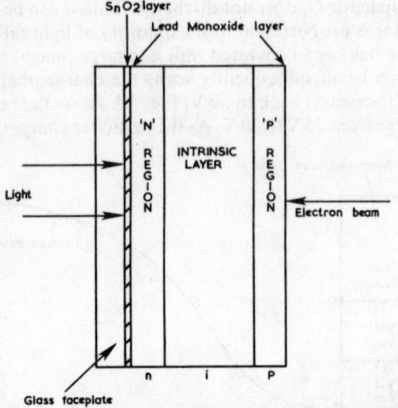

FIG. 3.7 DETAILS OF TARGET IN PLUMBICON TUBE

The signal plate consists of a thin transparent film of conductive tin oxide located between the glass faceplate and the n-region of the light sensitive layer. During operation the signal plate potential is held positive to the cathode as for the vidicon.

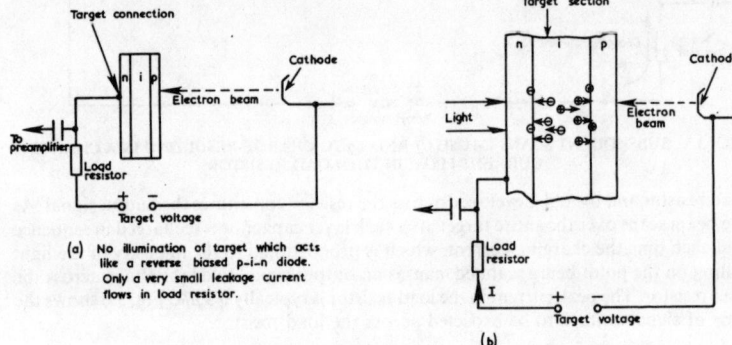

FIG. 3.8 DIAGRAMS ILLUSTRATING BASIC OPERATION OF PLUMBICON

The lead monoxide layer forms a p-i-n diode* (or many p-i-n diodes in parallel) and during operation is reversed biased. Thus with no light falling on the camera tube, only a very small leakage current (the dark current) flows in the p-i-n diode and load resistor, Fig. 3.8(a). The dark current is of the order of 0·003 μA which is much smaller than in the vidicon. When light falls on the lead monoxide layer, bonds are broken in the intrinsic layer releasing holes and electrons which find their way to the conductive 'n' and 'p' regions, Fig. 3.8(b). The number of charge carriers released is dependent on the intensity of the incident light. As a result of this action the potential of the 'p' region increases. When the electron beam scans the target layer the beam neutralises the charge until the 'p' region drops to cathode potential. This causes a current flow in the load resistor which produces the output signal as in the vidicon.

Apart from its low lag characteristic, the plumbicon has a uniform low level dark current and has a high sensitivity, i.e. is capable of working at low-light levels. The plumbicon tube is used in monochrome and colour cameras for broadcast work and in scientific applications.

*Silicon p-i-n switching diodes consist of a layer of intrinsic silicon (the i layer), sandwiched between heavily doped 'n' and 'p' material. Under reverse bias the i layer is almost devoid of carriers and only a small leakage current flows. With forward bias the i layer is filled with carriers. The device will, therefore, function as a switch.

The Output Signal

In a monochrome television system, the camera output signal should bear an amplitude proportional to the LUMINANCE of the coloured light falling on the camera lens. Strictly speaking, the response of the camera to different colours should closely follow the response of the human eye (see Fig. 1.7) and in practice is fairly similar. The colour response of a monochrome camera tube is entirely determined by the chemical composition and processing of the target layer.

In considering the monochrome camera output signal for various scenes it will be assumed that the camera produces a maximum output of 1 V which corresponds to maximum luminance and 0 V for zero luminance. Fig. 3.9(a) shows the luminance signal over the period of one line for two sample line periods *A–B* and *C–D*. Note that

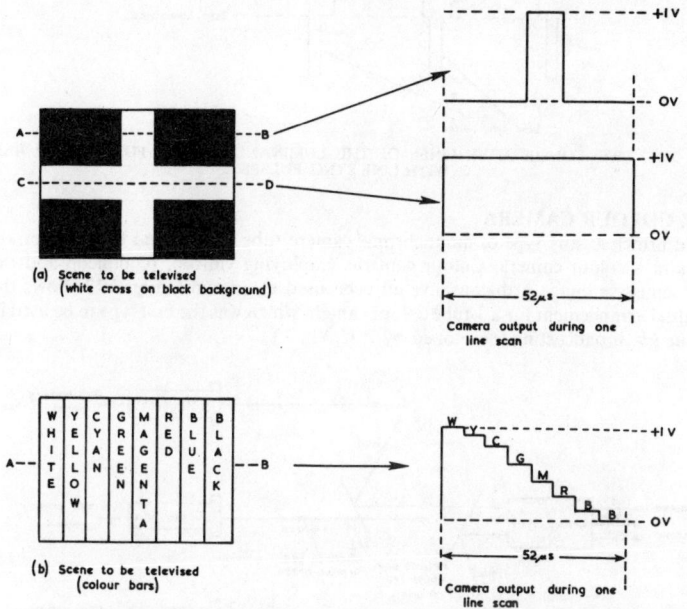

(a) Scene to be televised
(white cross on black background)

Camera output during one line scan

(b) Scene to be televised
(colour bars)

Camera output during one line scan

FIG. 3.9   CAMERA OUTPUT SIGNAL FOR TEST PICTURES (LUMINANCE SIGNAL E_Y)

since white has the highest luminance, the luminance signal level on white reaches 1 V and on black (zero luminance) falls to 0 V. For the colour bars shown at (b), the camera produces a staircase output waveform. The steps represent the decreasing order of luminance of the colours making up the bars of this test signal. The luminance signal waveform will be the same on all lines of the picture for this scene. The staircase waveform is often referred to as a 'grey-scale', when applied to a monochrome display tube or colour tube (with the colour control at minimum); the effect is a graduation of greys going from white on the left-hand side of the screen to black on the right-hand side.

Fig. 3.10† shows the luminance signal during 1 line scan corresponding to *A–B* for a normal type of scene. Note that although blue has a lower luminance than red, the sky being light blue (blue + white) will probably have a higher luminance value than the red mast. Of course, for this picture the luminance signal will be quite different for a line scan, say, passing through the hull of the boat. The luminance signal is normally designated the E_Y signal.

†In the fold-out colour section at the back of the book

Although in a monochrome system colours are reproduced in black and white, it is possible to distinguish between the flesh tones of white, brown and yellow skinned people because the camera is given a colour response similar to that of the human eye.

In practice line and field sync. pulses must be added to the E_y signal output of the camera. Fig. 3.11 shows two consecutive lines of luminance information for the colour bars of Fig. 3.9(b) together with line synchronising pulses.

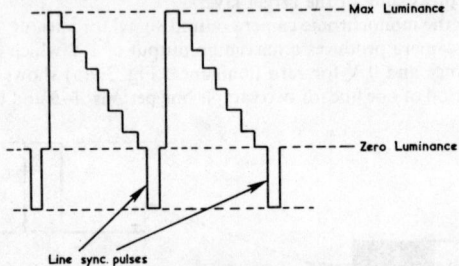

FIG. 3.11   TWO CONSECUTIVE LINES OF THE LUMINANCE SIGNAL FOR COLOUR BARS WITH LINE SYNC. PULSES

THE COLOUR CAMERA

In principle, any type of monochrome camera tube can be used for the separate tubes in a colour camera. Colour cameras employing vidicon, plumbicon and the photoemissive image orthicon have all been used at one time. Fig. 3.12 shows the essential arrangement for a 3-tube colour camera which was the first type to be used in colour t.v. broadcasting (developed by R.C.A.).

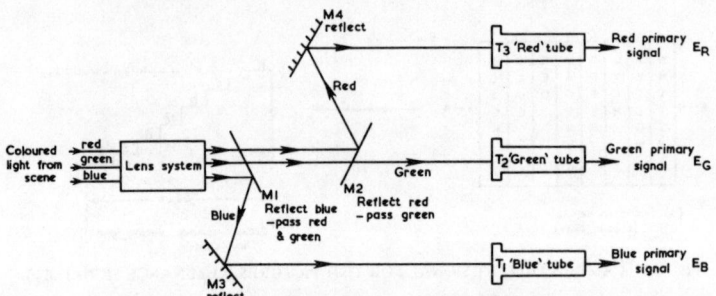

FIG. 3.12   BASIC ARRANGEMENT OF A 3-TUBE COLOUR CAMERA

The first process in a colour television system is to break down the coloured light coming from the scene into its basic components of red, green and blue light. In the 3-tube camera this is done with the aid of 'dichroic mirrors', which pass certain light wavelengths but reflect others. Light from the scene passes through a common lens system and falls on the dichroic mirror M_1 which passes red and green light but reflects the blue component. The blue light is reflected by a silvered surface mirror M_3 on to the face of T_1, the 'blue' camera tube. The red and green light components passing through M_1 fall on another dichroic mirror M_2 which passes the green light on to the face of T_2, the 'green' camera tube, but reflects the red component. M_4, another silvered surface mirror, then reflects the red light on to the face of T_3, the 'red' camera tube. Other correcting filters are normally required but are not essential to the basic idea of operation.

Each camera tube has its own electron beam and deflection system. The separate colour images formed on the sensitive target layer of the three tubes must be scanned in synchronism at line and field rate as for the basic camera. As the images are scanned, video signal voltages are produced SIMULTANEOUSLY at the output of each camera tube. These output signals are called 'primary signals' since they are related to the three primary colours and are designated E_R, E_G and E_B.

Dichroic Mirrors

Dichroic mirrors are made by the evaporation of very thin layers of materials on a glass sheet with alternate layers having high and low refractive indices. Each layer (there may be up to 20) is usually made a quarter of a wavelength thick at the rejection light frequency.

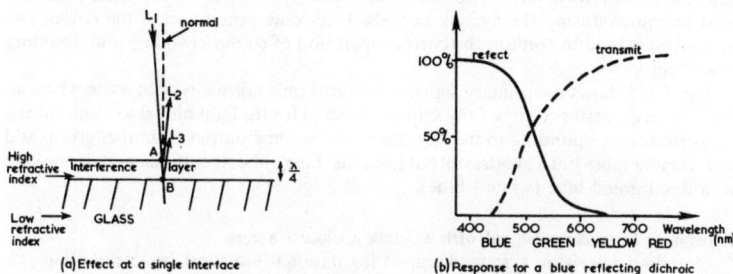

(a) Effect at a single interface   (b) Response for a blue reflecting dichroic

FIG. 3.13   PRINCIPLE OF DICHROIC MIRROR

Fig. 3.13(a) shows a single interface where an incident ray of light L_1 strikes the interference layer almost at right angles. At point A some of the light is reflected (L_2) and the rest passes through the interference layer. On reaching point B reflection again occurs (L_3). As the thickness of the layer is $\lambda/4$ of the incident light which strikes the mirror almost perpendicularly, L_2 and L_3 will be out of phase by 180° due to the extra $\lambda/2$ travelled by L_3. Also, when L_3 is reflected at the high-to-low refractive index interface there is a further 180° phase change, i.e. L_2 and L_3 are 360° out of phase or in phase. Thus L_2 and L_3 rays have the effect of reinforcing each other, i.e. at this particular incident light wavelength reflection occurs from the dichroic mirror. Clearly, this condition will apply for light of one specific wavelength, say, 760 nm. An incident light ray of half this wavelength (380 nm) would cause a 180° phase difference between L_2 and L_3 (as the interference layer would·be $\lambda/2$ thick to the incident light), thus no reflection would occur. Between these two extremes wavelengths there would be a changeover from reflecting to transmitting. It should be noted that if the light passes from a medium of low refractive index to a medium of higher refractive index there is no 180° phase change at the interface. Thus, for the wavelengths quoted the conditions would be reversed; the 760 nm reflected rays L_2 and L_3 would cancel but the 380 nm reflected rays L_2 and L_3 would add thus giving a blue reflecting dichroic mirror; Fig. 3.13(b).

Everyday examples of the dichroic effect are to be seen in the colours reflected from soap bubbles and thin films of oil on water.

Primary Signal Waveforms

We will now consider the primary signal waveshapes for the three sample scenes used previously. In Fig. 3.14(a)† the primary signal waveforms for a sample line scan corresponding to $A–B$ are shown. Note that on white all the camera tubes produce an output since the colour camera breaks down the white light into its three basic components of red, green and blue light. If the 'standard white light' (illuminant D) is used and is at maximum intensity, the camera tube outputs are each adjusted to give 1 V of signal.

The E_G, E_R and E_B outputs for the colour bars are shown in Fig. 3.14(b).† Note for example, that since green light is only present in the white, yellow, cyan and green

†In the fold-out colour section at the back· of the book

bars, the 'green' camera tube produces an output for these bars, but no output for the remaining bars. By similar reasoning the E_R and E_B output waveforms may be deduced. Of course, on the colour bars these waveforms will be repetitive line by line.

Instead of using a colour bar test card and colour camera, the primary signal waveforms for the colour bars may be generated electronically. With, say, a master square-wave oscillator working at 76·923 kHz, four complete cycles would be produced in the active line period of 52 μs and the oscillator output could be used for the E_B waveform. By dividing this output by a factor of 2, say, by employing a bistable oscillator, the E_R waveform may be produced and then with a further divide-by-two stage the E_G waveform may be obtained. Of course, the output of the master oscillator must be muted during the flyback periods. Electronic generation of the colour bar waveforms is used to confirm the correct operation of colour encoding and decoding equipment.

Fig. 3.15† shows the primary signal waveforms on a normal type of scene where all the colours are assumed to be fully saturated except for the light blue sky. Thus, during the periods corresponding to the sky there will be some output from the 'green' and 'red' camera tubes but a greater output from the 'blue' tube since the camera is 'looking at' a desaturated blue (white + blue).

Forming a Luminance Signal with a 3-tube Colour Camera

A colour television system designed for national use must be 'compatible', i.e. monochrome receivers should be able to tune into the colour transmission and extract the necessary information to produce a monochrome picture. The E_R, E_G and E_B signals contain hue, saturation and some luminance information. However, individually these signals do not contain sufficient luminance information to produce a good black-and-white picture on a monochrome receiver. The method adopted to produce a luminance signal with a 3-tube colour camera is shown in Fig. 3.16.

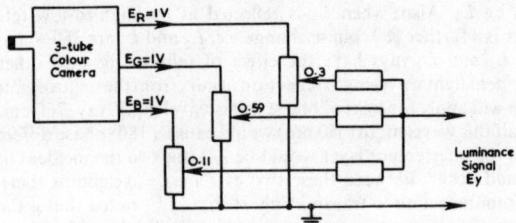

FIG. 3.16   FORMING THE LUMINANCE SIGNAL WHEN A 3-TUBE COLOUR CAMERA IS USED

The primary signal outputs from the camera video amplifiers are adjusted so that they are nominally 1 V on the standard white at maximum intensity. Portions of the primary signals are taken to form a luminance signal. The actual portions taken are $0.59E_G$, $0.3E_R$ and $0.11E_B$. When added together in the resistive matrix shown they give the luminance signal E_Y which can be written

$$E_Y = 0.3E_R + 0.59E_G + 0.11E_B.$$

The particular portions taken to form the luminance signal are related to the relative contribution to luminance of the colour phosphors used on the colour display tube.

Consider the formation of the luminance signal for the colour bars:

WHITE: $E_Y = 0.3 + 0.59 + 0.11 = 1$ V
YELLOW: $E_Y = 0.3 + 0.59 + 0.00 = 0.89$ V
CYAN: $E_Y = 0.00 + 0.59 + 0.11 = 0.7$ V
GREEN: $E_Y = 0.00 + 0.59 + 0.00 = 0.59$ V

†In the fold-out colour section at the back of the book

MAGENTA: $E_Y = 0.3 + 0.00 + 0.11 = 0.41$ V
RED: $E_Y = 0.3 + 0.00 + 0.00 = 0.3$ V
BLUE: $E_Y = 0.00 + 0.00 + 0.11 = 0.11$ V
BLACK: $E_Y = 0.00 + 0.00 + 0.00 = 0.00$ V

The luminance signal waveform during one line of the colour bars is shown in Fig. 3.17. As for the monochrome camera output [see Fig. 3.9(b)] it will be noted that the steps on the waveform are arranged in decreasing order of luminance from left to right.

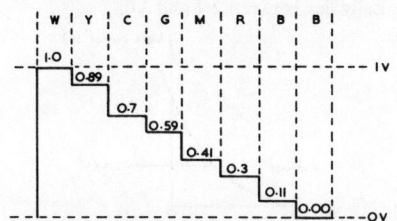

FIG. 3.17 LUMINANCE SIGNAL E_Y FOR COLOUR BARS

The 4-tube Colour Camera

A disadvantage of the 3-tube camera is that to obtain a luminance signal of adequate sharpness the separate images of the red, green and blue tubes must be very closely registered with one another. Also, the luminance signal is not entirely correct owing to gamma correction. To deal with these problems the 4-tube camera was designed, the extra tube being used to produce the luminance signal.

Fig. 3.18 shows one arrangement which is basically a combination of a 3-tube colour camera and a monochrome camera. Light coming from the lens system falls on

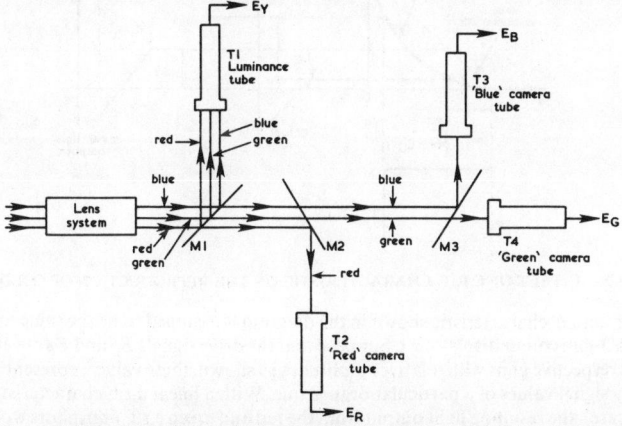

FIG. 3.18 A 4-TUBE COLOUR CAMERA (SEPARATE TUBE USED FOR THE LUMINANCE SIGNAL)

M_1, a semi-silvered mirror, and reflects part of the light on to the face of T_1, the luminance tube. This tube is given a spectral response to approximately match the response of the human eye and provides at its output a separate luminance signal at high definition. The light passing through M_1 is split into its three basic colours by the action of the dichroic mirrors M_2 and M_3. Thus, as for the 3-tube camera, separate colour images are produced on T_2, T_3 and T_4 which provide the primary signals.

Plumbicon tubes may be used for all four tubes, but with modern cameras prisms with dichroic surfaces are used in place of the dichroic mirrors, so providing a more compact optical system.

Gamma Correction

If a television system were linear in operation gamma correction would not be required. In practice, the output of a t.v. camera usually has to be 'gamma corrected' because of the characteristic of the display c.r.t. A typical relationship between the light output (L) and the grid signal drive voltage (V) is shown in Fig. 3.19. The light output is related to the drive voltage by the following expression:

$$L = kV^\gamma$$

where k is a constant and γ (gamma) another constant which typically lies between 2·7 and 3·0.

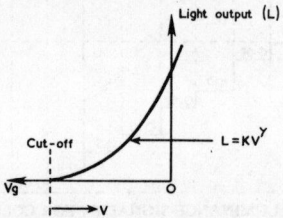

FIG. 3.19   TYPICAL C.R.T. CHARACTERISTIC

Without any compensation for this non-linearity in the c.r.t. characteristic, the luminance signal and colour rendering will be incorrect. The effect on colour reproduction is shown in Fig. 3.20.

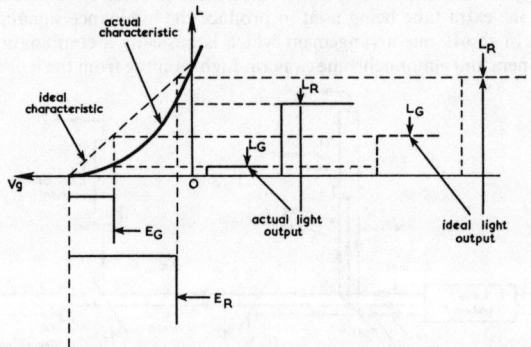

FIG. 3.20   EFFECT OF C.R.T. CHARACTERISTIC ON THE REPRODUCTION OF COLOUR

The 'actual' characteristic shown in the diagram is assumed to be the same for each gun of a 3-gun colour display c.r.t. Suppose that the drive signals E_G and E_R are applied to their respective guns with relative amplitudes as shown, these values representing the primary signal values of a particular orange hue. With a linear c.r.t. characteristic (the 'ideal' case), the resulting light output from the red and green c.r.t. phosphors would be as shown dotted and the correct orange hue would be displayed. However, because of the non-linearity the actual light outputs will be as shown by the solid lines with the red light output being less than intended but with the green light diminished even more. As a result the reproduced hue would be a deeper orange, i.e. the original hue has been distorted.

It is most convenient to correct for the c.r.t. non-linearity at the camera end of the system because the camera itself is also non-linear in operation and this affects the amount of correction required. The correction is carried out in a gamma correcting amplifier. This amplifier is arranged to have the opposite law to that of the c.r.t. (assuming the camera to be linear in operation). Consider a single primary signal voltage, say, E_R applied to the input of the amplifier, Fig. 3.21. The output signal

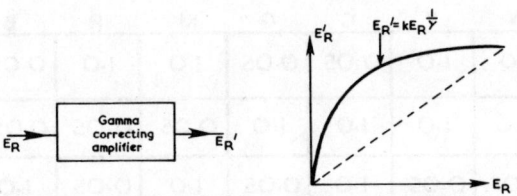

FIG. 3.21   GAMMA CORRECTION AMPLIFIER AND ITS TRANSFER CHARACTERISTIC

E_R'(the prime ' signifying that gamma correction has taken place) is related to the input signal E_R by the following:

$$E_R' = kE_R^{\frac{1}{\gamma}}$$

where γ is the gamma of the c.r.t.

As shown, the correcting curve increases the amplifier output signal for low input signal levels, with the amount of correction decreasing with input signal amplitude. This overcomes the non-linearity of the c.r.t. resulting in a linear television system. With a 4-tube colour camera, the E_Y, E_R, E_G, and E_B signals are individually gamma-corrected to give E_Y', E_R', E_G' and E_B' at the correcting amplifier outputs.

A table for 100% amplitude, 100% saturated colour bars giving the values of the primary and luminance signals is shown in Fig. 3.22. These values will be used throughout the remainder of the book when dealing with colour-difference signals and

	W	Y	C	G	M	R	B	B
E_R	1·0	1·0	0·00	0·00	1·0	1·0	0·00	0·00
E_G	1·0	1·0	1·0	1·0	0·00	0·00	0·00	0·00
E_B	1·0	0·00	1·0	0·00	1·0	0·00	1·00	0·00
E_Y	1·0	0·89	0·7	0·59	0·41	0·3	0·11	0·00

FIG. 3.22 TABLE OF VALUES OF PRIMARY AND LUMINANCE SIGNALS FOR 100% SATURATED COLOUR BARS

phasor diagrams (which are considered in the next chapter). In practice, the transmitted colour bars are 100% amplitude, 95% saturated and the effect on the signal values is shown in Fig. 3.23. This table also gives the gamma-corrected signal values (using an assumed c.r.t. gamma of 2·8). The E_Y' signal is that produced by matrixing the gamma-corrected outputs from a 3-tube colour camera.

	W	Y	C	G	M	R	B	B
E_R	1.0	1.0	0.05	0.05	1.0	1.0	0.05	0.00
E_G	1.0	1.0	1.0	1.0	0.05	0.05	0.05	0.00
E_B	1.0	0.05	1.0	0.05	1.0	0.05	1.0	0.00
E_Y	1.0	0.89	0.72	0.6	0.44	0.33	0.16	0.00
E'_R	1.0	1.0	0.34	0.34	1.0	1.0	0.34	0.00
E'_G	1.0	1.0	1.0	1.0	0.34	0.34	0.34	0.00
E'_B	1.0	0.34	1.0	0.34	1.0	0.34	1.0	0.00
E'_Y	1.0	0.93	0.8	0.73	0.61	0.54	0.41	0.11

FIG. 3.23 TABLE OF VALUES OF PRIMARY AND LUMINANCE SIGNALS (UNCORRECTED AND GAMMA CORRECTED) FOR 100% AMPLITUDE AND 95% SATURATED COLOUR BARS (gamma = 2·8)

CHAPTER 4

ENCODING THE COLOUR SIGNAL

I t was seen in Chapter 2 that for satisfactory monochrome operation the monochrome or luminance signal requires a video bandwidth extending up to 5·5 MHz. During a colour t.v. broadcast, extra information called the 'chrominance' or 'colouring' signal has to be transmitted. Because extra channels or bandwidth could not be spared to accommodate the chrominance signal, it became vitally important in evolving a colour t.v. system for national use to find a way of fitting the chrominance signal information within the normal monochrome video bandwidth. At first sight a seemingly insoluble problem but as will be shown later it was cleverly resolved. Another important consideration is that the signals sent out during a colour t.v. broadcast must be composed in such a way that a monochrome receiver is able to pick up a suitable signal in order to produce a good monochrome picture. This feature is called 'compatibility'. Also, as all transmissions are not in colour, the colour t.v. system must have 'reverse compatibility' so that a colour receiver can display a good quality monochrome picture.

So far we have seen that there are four signals available at the output of a 4-tube colour camera, Fig. 4.1. With a 3-tube camera, the luminance signal is produced by matrixing the primary signals; thus at the output of the matrixing network all four

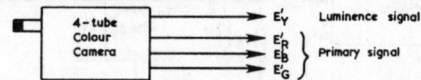

FIG. 4.1 THE SIGNAL OUTPUTS OF A 4-TUBE COLOUR CAMERA

signals are available as for the 4-tube camera. To satisfy the requirements mentioned above the following signals are transmitted:

(i) E_Y' luminance signal

(ii) $\left. \begin{array}{l} E_R' - E_Y' \\ E_B' - E_Y' \end{array} \right\}$ colour-difference signals.

(i)   This signal is transmitted at wide bandwidth (0—5·5 MHz) and provides the necessary information to produce a good monochrome image on black-and-white and colour receivers.

(ii) These signals are transmitted at small bandwidth (0—1 MHz) and together with the signal in (a) provide all the necessary information to produce a well defined colour image on a colour receiver.

Although in a colour t.v. system we start off at the camera end by generating primary signals which are eventually required in the receiver to drive the colour display tube, these signals are not transmitted directly for the following reasons.

(a)   As previously noted the primary signals contain hue, saturation and some luminance information. The luminance signal is created to carry all the luminance information, so it would be very inefficient to duplicate this information by sending out primary signals.

(b)   Using the colour-difference signal method we need only transmit two signals extra to the luminance signal thus making the task of fitting the chrominance information within the normal monochrome bandwidth an easier one.

To drive a colour display tube at the receiver the E_R', E_B' and E_G' signals are required. However, if the colour-difference and luminance signals are available at the

37

receiver the primary signals may be recovered as can be seen from the following simple algebraic addition:

$$(E'_R - E'_Y) + E'_Y = E'_R$$
$$(E'_B - E'_Y) + E'_Y = E'_B$$

Also, if $E'_G - E'_Y$ is available as well at the receiver

$$(E'_G - E'_Y) + E'_Y = E'_G.$$

It will be noted that an $E'_G - E'_Y$ colour-difference signal is not transmitted, but this too can be recovered at the receiver by adding together suitable portions of the $E'_R - E'_Y$ and $E'_B - E'_Y$ signals, which will be explained later.

COLOUR-DIFFERENCE SIGNALS

Fig. 4.2 shows the principle involved in the production of the colour-difference signals. A 3-tube colour camera has been used in this diagram as it illustrates more clearly some basic relationships between the various signals. Gamma-corrected

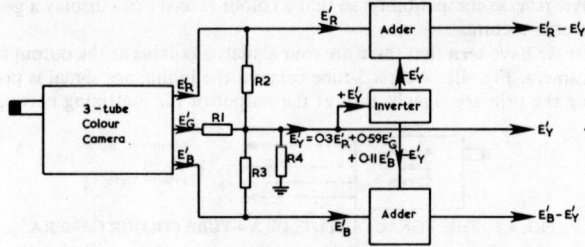

FIG. 4.2   PRODUCING THE COLOUR-DIFFERENCE SIGNALS

primary signals available from the output of the camera are fed to the resistive matrixing network comprising $R_1 - R_4$ to produce the gamma-corrected luminance signal, as previously explained. In practice a resistive matrix is not used owing to a number of disadvantages. More usually, the signals are attenuated and then combined *via* three transistors feeding a common load. The output of the matrix, designated $+ E'_Y$ is fed to an inverter which provides $- E'_Y$ at its output. The signal is fed to the two adders where it is added to E'_R and E'_B. As a result of the addition, colour-difference signals are produced at the outputs of the adder units.

It should be noted that a colour-difference signal is simply the subtraction of electrical voltages and has nothing to do with coloured light. The result of the subtraction of the luminance signal voltage from the primary signal voltages for the colour bars are shown in Fig. 4.3. The corresponding video signal waveforms are given in Fig. 4.4.

The merits of the colour-difference signal method are:

(i)   When only monochrome is being transmitted, the colour-difference signals automatically disappear and no chrominance is transmitted. Thus no colour noise will appear on the screen of the colour display tube from the chrominance channel of the receiver.

(ii)   The colour-difference signals are true colour signals, *i.e.* they carry hue and saturation information only, the luminance information being carried by the luminance signal. This is referred to as the 'constant luminance' principle which means that the luminance of the light emanating from the screens of a monochrome receiver and a colour receiver is exactly the same when both are tuned to the same colour transmission, although one picture is in monochrome and the other in colour. In practice, constant luminance is not fully met owing to gamma correction.

	W	Y	C	G	M	R	B	B
E_R	1.0	1.0	0.00	0.00	1.0	1.0	0.00	0.00
E_B	1.0	0.00	1.0	0.00	1.0	0.00	1.0	0.00
E_G	1.0	1.0	1.0	1.0	0.00	0.00	0.00	0.00
E_Y	1.0	0.89	0.7	0.59	0.41	0.3	0.11	0.00
E_R-E_Y	0.00	+0.11	−0.7	−0.59	+0.59	+0.7	−0.11	0.00
E_B-E_Y	0.00	−0.89	+0.3	−0.59	+0.59	−0.3	+0.89	0.00
E_G-E_Y	0.00	+0.11	+0.3	+0.41	−0.41	−0.3	−0.11	0.00

FIG. 4.3 TABLE SHOWING COLOUR-DIFFERENCE SIGNAL VALUES FOR 100% AMP., 100% SAT. COLOUR BARS

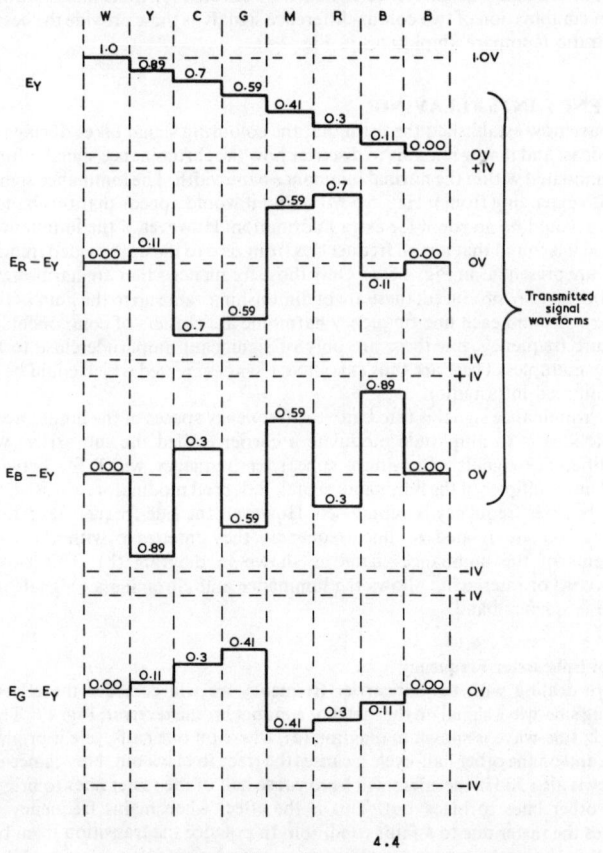

4.4

FIG. 4.4 COLOUR-DIFFERENCE SIGNAL WAVEFORMS

$E'_G - E'_Y$ **Signal**

There is no need to transmit the $E'_G - E'_Y$ signal because of a fundamental relationship that exists between the three colour-difference signals. This may be seen from the following:

$$\text{Now } E'_Y = 0.3E'_R + 0.59E'_G + 0.11E'_B \tag{4.1}$$
$$\text{and } E'_Y = 0.3E'_Y + 0.59E'_Y + 0.11E'_Y \tag{4.2}$$

Subtracting **(4.2)** from **(4.1)** we have

$$O = (0.3E'_R - 0.3E'_Y) + (0.59E'_G - 0.59E'_Y) + (0.11E'_B - 0.11E'_Y)$$

$$\text{or } O = 0.3(E'_R - E'_Y) + 0.59(E'_G - E'_Y) + 0.11(E'_B - E'_Y)$$

$$\text{Thus } 0.59(E'_G - E'_Y) = -0.3(E'_R - E'_Y) - 0.11(E'_B - E'_Y)$$

Dividing both sides by 0.59

$$(E'_G - E'_Y) = -0.51(E'_R - E'_Y) - 0.19(E'_B - E'_Y).$$

Thus if the red and blue difference signals are available at the receiver, all that is required is to add -0.51 of $E'_R - E'_Y$ to -0.19 of $E'_B - E'_Y$ to obtain the green signal. The minus sign simply indicates a phase reversal of the red and blue difference-signals. The red and blue difference-signals were selected for transmission rather than any other combination of two colour-difference signals as these provide the best signal-to-noise ratio (compare amplitudes in Fig. 4.4).

FREQUENCY INTERLEAVING

We have now established the form that the colouring signal takes during a colour t.v. broadcast and it now remains to discover how the chrominance signal information is accommodated within the normal luminance bandwidth. The luminance signal has a bandwidth extending from 0 Hz—5·5 MHz and it would appear that this band is full, thus there would be no room for extra information. However, if the luminance signal is analysed it is found that not all frequencies from zero to the upper video frequency of 5·5 MHz are present as in Fig. 4.5(a). Only those frequencies that are harmonics of the LINE FREQUENCY are produced; these are of diminishing value up to the limit of the band (5·5 MHz). Around each line frequency harmonic are clusters of components at field and picture frequency, but these are only of significant amplitude close to the line frequency multiples. There are thus FREQUENCY SPACES created which could be used to carry additional information.

The chrominance signal is fitted into the frequency spaces of the luminance signal. The basic idea is to amplitude modulate a carrier (called the subcarrier) with the colour-difference signals, choosing a subcarrier frequency which fits between two adjacent line multiples of the luminance signal. Balanced modulators are used, thus the actual subcarrier frequency is suppressed. However, the side-frequencies remain and since they too are related to line frequency, they interleave with the harmonic components of the luminance signal as shown in diagram (b). This FREQUENCY INTERLEAVING or interlacing allows the luminance and chrominance signals to share the same frequency band.

Choice of Subcarrier Frequency

Before dealing with the subcarrier frequency we will consider the effect of an interfering sine-wave signal on the raster of a monochrome receiver, Fig. 4.6. The effect of a 50 Hz sine-wave is shown in diagram (a), where on one half-cycle it brightens up the trace and on the other half-cycle it causes the trace to black out. Now, since the field frequency is also 50 Hz the effect will be to cause half of the raster lines to brighten up and the other lines to black out. This is the effect when mains frequency voltage modulates the raster due to a fault condition. In practice the transition from black to white will not be as sharp as indicated since we are assuming sine-wave modulation as

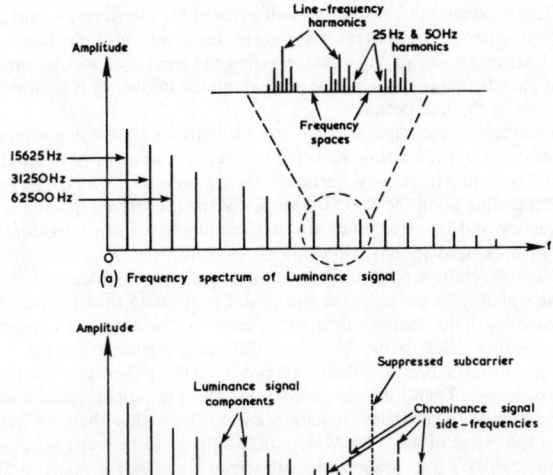

(a) Frequency spectrum of Luminance signal

(b) Frequency Interleaving of Luminance and Chrominance signals

FIG. 4.5 FITTING THE CHROMINANCE INFORMATION INTO THE LUMINANCE SIGNAL BANDWIDTH

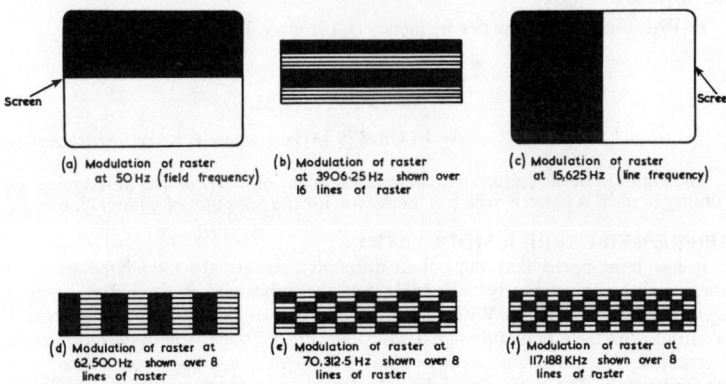

(a) Modulation of raster at 50 Hz (field frequency)

(b) Modulation of raster at 3906·25 Hz shown over 16 lines of raster

(c) Modulation of raster at 15,625 Hz (line frequency)

(d) Modulation of raster at 62,500 Hz shown over 8 lines of raster

(e) Modulation of raster at 70,312·5 Hz shown over 8 lines of raster

(f) Modulation of raster at 117·188 KHz shown over 8 lines of raster

FIG. 4.6 EFFECT OF SINE-WAVE MODULATION OF RASTER AT VARIOUS FREQUENCIES ON MONOCHROME RECEIVER SCREEN

opposed to square-wave. Diagram (b) shows the results when the periodic time of the interfering signal is exactly four times the period of one line scan. In this case the signal will brighten up the trace for two consecutive line scans but black out the following two lines. Due to interlacing, however, the effect on the screen will be four line scans blacked out followed by four line scans brightened up, providing a series of horizontal bars. When the frequency of the interfering signal is at line frequency, diagram (c), half of each line scan brightens up and the other half blacks out creating the same effect as at (a) but rotated through 90° *i.e.* vertical bars. As the frequency increases so does the number of vertical bars, diagram (d), where the interfering signal is exactly four times the line frequency. If the interfering signal is an exact multiple of the line frequency, vertical bars are produced. However, if the signal is, say, 4·5 times line frequency the effect is to break up the vertical bars into a series of dots lying at an angle of 45°, diagram (e). This is because there is an odd half-cycle of the interfering signal in the line period. However, when the pattern is broken up in this way its visibility is considerably reduced thus lessening its annoying effect. Raising the frequency further produces an even finer dot pattern, diagram (f), where once again the frequency is such that an odd half-cycle occurs in the line period.

The main features of the effect of interfering signals on a television screen are: (a) When the interfering signal is below half-line frequency a series of horizontal bars are produced. (b) Above line frequency, vertical bars are produced which may be broken up into dots depending upon the exact frequency of the interfering signal. (c) Between half-line frequency and line frequency the pattern changes from horizontal bars to vertical bars by making diagonal stripes on the screen.

As far as a monochrome receiver is concerned, the chrominance signal sidebands are interfering signals and as such give rise to dot patterning of the raster lines. The higher the frequency of the chosen subcarrier, the finer is the dot pattern produced and the less its annoyance value. Now, the colour-difference signals are in the range 0—1 MHz thus the chrominance sidebands extend 1 MHz, either side of the chosen subcarrier frequency. Therefore, to accommodate the upper side-band of the chrominance signal the subcarrier frequency cannot be higher than 4·5 MHz. Thus somewhere in the range of 4·0—4·5 MHz would appear to be most suitable.

In the American N.T.S.C. system, the subcarrier frequency is made to fit exactly between two adjacent line harmonics of the luminance signal. This is known as 'half-line offset' and results in a dot pattern of minimum visibility. With the PAL system (used in the U.K.) half-line offset is not used because of the phase reversal of the V signal (see later). Instead, 'quarter-line offset' is employed, *i.e.* the subcarrier frequency is so many and a quarter times the line frequency or so many and three-quarters times line frequency.

In PAL the exact subcarrier frequency (f_{sc}) is given by:

$$f_{sc} = f_L(284 - \tfrac{1}{4}) + 25 \text{ Hz}$$
$$= 15625 \times 283·75 + 25 \text{ Hz}$$
$$= 4·43361875 \text{ MHz} \quad \text{where } f_L \text{ is the line frequency.}$$

The addition of the picture frequency component of 25 Hz assists in reducing the visibility of the dot pattern, which is the reason for the adoption of quarter-line offset.

SUPPRESSED-CARRIER MODULATION

It has been noted that the colour-difference signals are modulated on to a subcarrier of approximately 4·43 MHz but the subcarrier is suppressed, thus a balanced modulator is used. With a suppressed-carrier or balanced modulator there is no output when the modulating signal is zero, see Fig. 4.7 where sine-wave modulation is assumed. When the modulating signal is present the output consists of side-frequencies only. The amplitude of the output varies in accordance with the amplitude of the modulating signal, but note that *each time the modulating signal changes polarity, the phase of the output changes by 180°*, see diagram (ii). Clearly, if the waveform of diagram (ii) were fed to an ordinary envelope detector, the detected output would be a very distorted version of the original modulation, because the carrier is 'missing'. Thus to reconstitute the original modulation, the carrier must be made available at the receiver when suppressed-carrier operation is employed. Diagram (iii) shows the output from an ordinary amplitude modulator for comparison with diagram (ii).

In practice the modulating signal is not a sine-wave but a colour-difference signal. An example of the modulator output when receiving $E_R - E_Y$ modulation (for the colour bars) is given in Fig.4.8. The important points to note are:

(a)   The output has an amplitude proportional to the amplitude of the colour-difference signal, *i.e.* no output on monochrome.

(b)   A 180° phase reversal occurs each time the colour-difference signal changes polarity

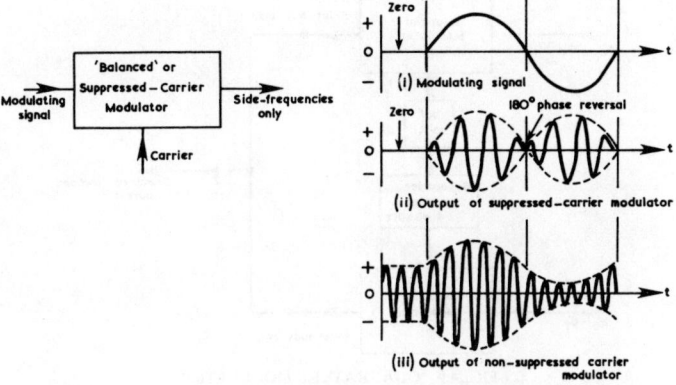

FIG. 4.7   SUPPRESSED CARRIER MODULATION

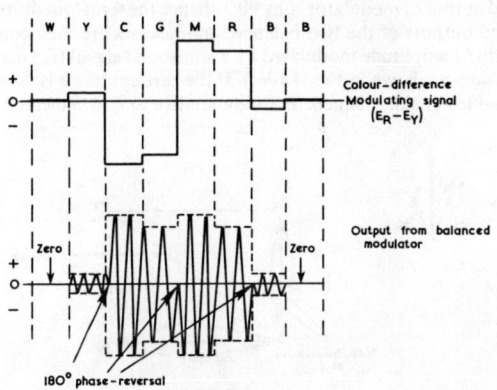

FIG. 4.8   EXAMPLE OF BALANCED MODULATOR OUTPUT WITH COLOUR-DIFFERENCE
MODULATING SIGNAL

Suppressed-carrier modulation is used in preference to ordinary amplitude modulation since with the former there is no large amplitude carrier to cause interference with monochrome reception. The side-frequencies, of course, produce some patterning but this is not so objectionable and is at its worst only on fully saturated colours.

QUADRATURE MODULATION

It has been explained how the chrominance and luminance information is fitted into the same frequency band using frequency interlacing. Now, the colouring information is carried by TWO colour-difference signals. Thus the problem is how to simultaneously modulate a common subcarrier with two colour-difference signals and to be able to separate them at the receiver. The problem is resolved by what is known as 'quadrature modulation'.

The basic arrangement for quadrature modulation is shown in Fig. 4.9. In balanced modulator A the blue difference-signal modulates a 4·43 MHz subcarrier, thus at x the output consists of side-frequencies only. Meanwhile, in balanced modulator B, the red difference-signal modulates a subcarrier having precisely the same frequency but

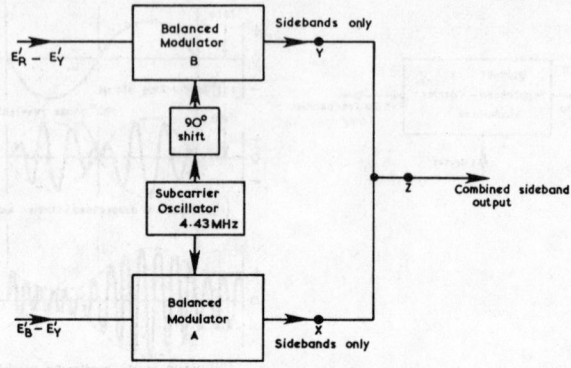

FIG. 4.9   QUADRATURE MODULATION

differing in phase by 90° to the subcarrier input to modulator A. The output of modulator B will also consist of side-frequencies only but these will differ in phase to the sideband output of modulator A by 90°—hence the term 'quadrature modulation'. The sideband outputs of the two balanced modulators are then combined.

A carrier (f_c) amplitude modulated by a sinusoidal signal (f_m) may be represented by three phasors as shown in Fig. 4.10(a). If the carrier phasor is assumed stationary, the upper and lower side-frequencies rotate relative to f_c as indicated with an angular

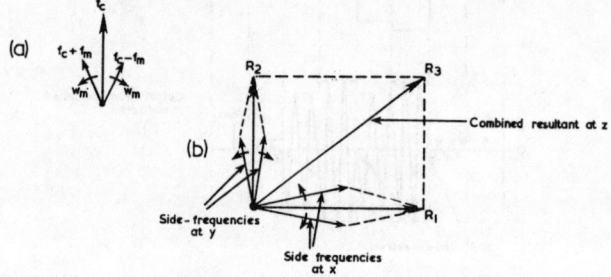

FIG. 4.10   COMBINING THE MODULATOR OUTPUTS

frequency ω_m. At any instant, the resultant is the sum of the three phasors. With suppressed-carrier operation the carrier phasor is omitted, leaving the side-frequency phasors. This representation is used in diagram (b) which shows the side-frequency output of modulator A together with the side-frequency output of modulator B. The resultant of the side-frequency output at x is the phasor R_1 whilst the resultant of the side-frequency output at y is the phasor R_2. When the outputs of the two modulators are combined the resultant is given by phasor R_3. This combined resultant represents a particular hue and saturation as decided by the respective amplitudes and polarities of the colour-difference signals. When dealing with the operation of a balanced modulator it was noted that the phase of the output changed by 180° whenever the modulating changed polarity. Since colour-difference signals may have either positive of negative polarity, the phase of the combined resultant may lie in any of the four quadrants of Fig. 4.11. The actual phase position of the combined resultant in any quadrant is dependent upon the relative amplitudes of the side-frequencies but these are directly proportional to the amplitudes of the colour-difference signals. Thus the colour-difference values may be plotted directly on the axes.

The phase position for the hues of the colour bars are shown in Fig. 4.12. The position of each phasor is obtained by taking the resultant of the respective values of

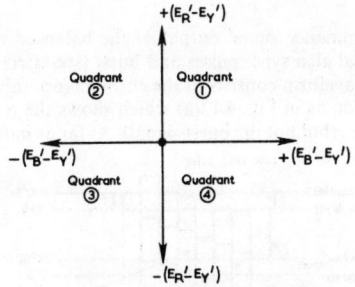

FIG. 4.11 MODULATION AXES

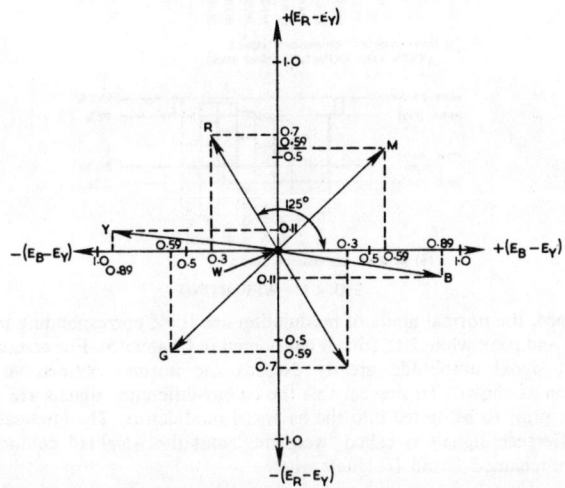

FIG 4.12 PHASE POSITIONS FOR THE COLOUR BARS 100% AMP., 100% SAT. (Non-weighted and without gamma correction)

$E_R - E_Y$ and $E_B - E_Y$ (see Fig. 4.3) for the various hues. The following points may be noted from this diagram:

(a) The HUE of a colour is determined by the PHASE POSITION of the resultant phasor. In angular measurement this is taken from the reference axis $+(E_B - E_Y) = 0°$, e.g. the red phasor makes an angle of approximately 125° with respect to the $+(E_B - E_Y)$ axis.

(b) The LENGTH of the resultant phasor is determined by the SATURATION OF A COLOUR AND ITS AMPLITUDE. The more desaturated a colour becomes the less will be the phasor's length. A fully desaturated hue (white) has a position at the origin.

(c) Complementary colours are diametrically opposite their associated primary colours, e.g. magenta is opposite green.

(d) If the phasor of a complementary colour is added to the phasor of its associated primary colour, the result is zero, i.e. white which is in accordance with the additive mixing of a primary colour and its complementary hue.

(e) By projecting from a particular phasor to the $(E_R - E_Y)$ and $(E_B - E_Y)$ axes the relative amplitudes of the resultants of the side-frequency components may be found.

In practice, Fig. 4.12 has to be modified on account of gamma correction and also because of 'weighting' which will be considered next.

WEIGHTING

The combined chrominance signal output of the balanced modulators together with the luminance signal also sync. pulses and burst (see later) are fed to the main u.h.f. modulator. The waveform consists of the chrominance information 'riding' on the luminance modulation as in Fig. 4.13(a) which shows the results on colour bars including line sync. pulses (but not the burst signal). As far as monochrome operation

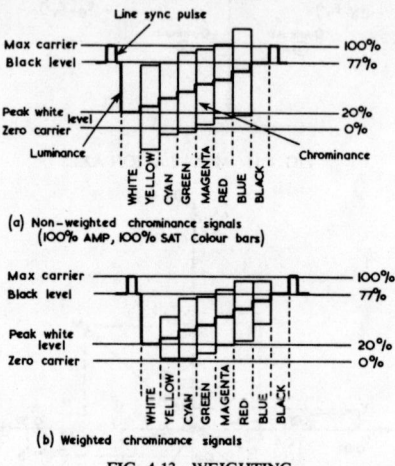

FIG. 4.13 WEIGHTING

is concerned, the normal limits of modulation are 100% corresponding to the sync. pulse tips and peak white 20% (this is considered in Chapter 5). For certain hues, the combined signal amplitude greatly exceeds the normal permissible limits of modulation as shown. To prevent this the colour-difference signals are reduced in amplitude prior to being fed into the balanced modulators. The attenuation of the colour-difference signals is called 'weighting' and the weighted colour-difference signals are renamed V and U signals where

$$V = 0.877 (E'_R - E'_Y)$$
$$\text{and } U = 0.493 (E'_B - E'_Y).$$

After weighting, the combined luminance and chrominance signal waveform is as in Fig. 4.13(b). It will be noted that the combined waveform no longer exceeds the tips of the sync. pulses nor does it reach zero carrier level. The weighting process does not cause any difficulty in the receiver as it is a simple matter to restore the colour-difference signals to their correct levels by adjusting the gains of the appropriate channels (called 'de-weighting').

From now on we shall be concerned mainly with V and U signals rather than $E'_R - E'_Y$ and $E'_B - E'_Y$ signals. The weighting process is shown in Fig. 4.14(a) where the inputs to the balanced modulators are now V and U signals or attenuated colour-difference signals. Because of this, the modulation axes are redesignated the V and U axes as shown in diagram (b).

THE PAL SYSTEM

A disadvantage of quadrature modulation is that since any hue is represented by a particular phase of the combined resultant output of the two balanced modulators, any phase shift of the resultant with respect to its correct phase position can cause wrong colours to be displayed at the receiver.

For example, consider the transmission of a particular red hue which is represented by the phasor R_t in Fig. 4.15. This signal has a true V component v_t and a

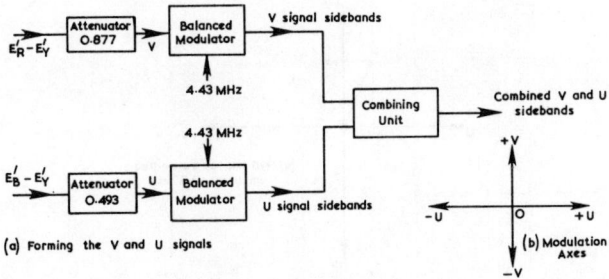

(a) Forming the V and U signals

(b) Modulation Axes

FIG. 4.14 V AND U SIGNALS

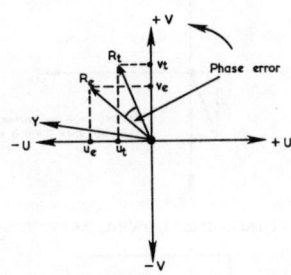

FIG. 4.15 EFFECT OF PHASE ERROR IN TRANSMITTED SIGNAL

true U component of u_t. Suppose that somewhere in the transmitting or receiving link a phase advance occurs causing the signal to take up the new phase position as shown. We will now call the phasor R_e as it represents an error signal. R_e has a V component v_e which is smaller than intended and a U component u_e which is larger than intended. After these components have been detected at the receiver, the display would show an incorrect hue (somewhere between red and yellow). Unfortunately the eye is very sensitive to variations in hue (particularly on flesh tones) and it is generally considered that the eye can detect hue variations caused by phase errors of about 5°.

Phase errors would not be such a serious matter if any phase shift occurring was the SAME for all transmitted hues and the colour burst. In this case a control could be fitted in the receiver to adjust the phase of the receiver subcarrier oscillator (this generates the subcarrier which was suppressed at the transmitter and which is essential for demodulation of the chrominance signals). There is, however, a type of phase shift which is level dependent, i.e. the amount of phase shift introduced depends upon the level of the chrominance signal. Note that the chrominance signal 'rides' on a luminance signal of varying level (see Fig. 4.13). With differing amounts of phase error for various colours it would be impossible to correct all the colours simultaneously with the aid of a single receiver control. Although level dependent or differential phase error may be reduced by careful design it cannot be wholly eliminated. Because of the problems associated with phase errors a modification to quadrature modulation was introduced called PAL (phase alternation line).

The basic idea of PAL is to invert the V component of the transmitted chrominance signal on alternate lines. Fig. 4.16(a) shows as an example a red hue having a $+v$ component and a $-u$ component. On odd lines the phasor for red is sent out as shown. On even lines a red hue is transmitted as in diagram (b). The phasor now has a $-v$ component but the U component is unaltered. Thus the transmitted phasor on even lines is a mirror image (about the U axis) of the transmitted phasor for odd lines. This simple modification overcomes the effects of phase errors. Of course, on even lines the

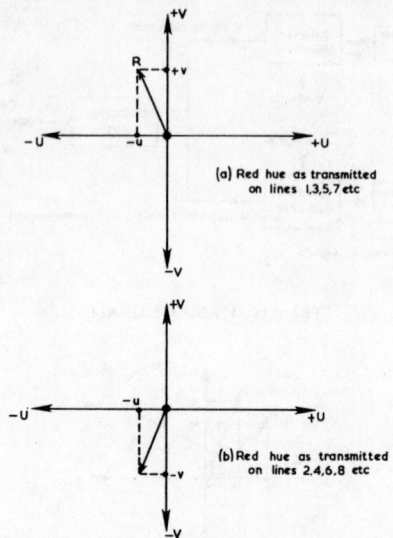

FIG. 4.16   DIAGRAMS SHOWING BASIC IDEA OF PAL

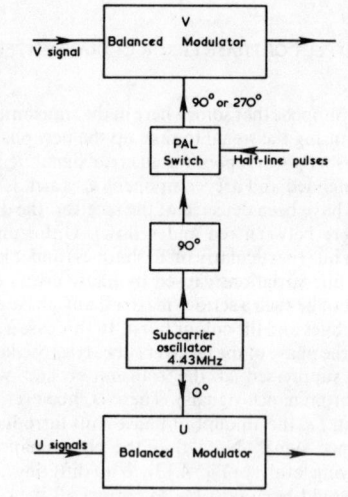

FIG. 4.17   INVERTING THE V SIGNAL COMPONENT

received phasor is incorrect so at the receiver the V component must be reinverted to put matters right.

Fig. 4.17 shows the arrangement used at the transmitter for inverting the V component on alternate lines. The subcarrier of 4·43 MHz is fed to the balanced modulators as previously described, but into the subcarrier path at the input to the V modulator is placed the inverting switch called the PAL switch. This switch is driven at half-line rate (7·8125 kHz) and on odd lines the subcarrier input to the V modulator is advanced by the normal 90° on the subcarrier input to the U modulator (0°). On even lines the PAL switch inverts the subcarrier so that it is at 270 with respect to the U

modulator subcarrier input. When the subcarrier input to the V modulator changes phase by 180 so does the side-frequency output.

Receiving PAL Signals

We will now consider the essential requirements of a colour receiver when dealing with PAL signals. Phasors for normal operation, *i.e.* no phase error in the received signal, are shown in Fig. 4.18. Here we are considering the transmission of a red hue, assumed to be sent out on all lines. Diagrams (a) and (b) show the transmitted and

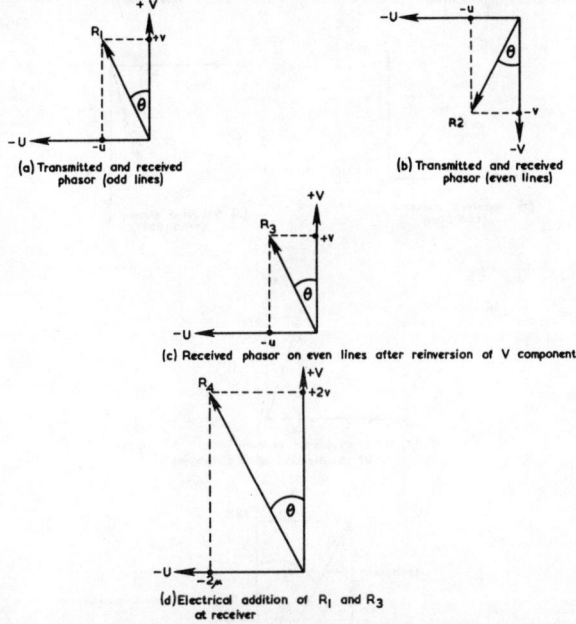

(a) Transmitted and received phasor (odd lines)

(b) Transmitted and received phasor (even lines)

(c) Received phasor on even lines after reinversion of V component

(d) Electrical addition of R_1 and R_3 at receiver

FIG. 4.18 DIAGRAMS SHOWING EFFECT OF PAL WHEN THERE IS NO PHASE ERROR (NORMAL OPERATION)

received phasor R_1 and R_2 for the red hue on odd and even lines. Now the phase of the received signal on even lines is incorrect, thus it is necessary *in the receiver to reinvert the V component* of R_2. This is shown in diagram (c) where after the V component has been reinverted we have the correct phasor (R_3) for the red hue.

In a PAL-D receiver (D stands for delay-line), the phasors of the odd and the corrected even lines are added together electrically and the resultant signal, after suitable processing, is used to produce colour on the display c.r.t. This is shown in diagram (d) where phasors R_1 and R_3 have been added vectorally to produce a resultant R_4. Note that R_4 makes the same angle θ with the $+V$ axis as R_1 and R_3, *i.e.* the hue is correct but is twice the length of R_1 or R_3. The increase in amplitude of the V and U components, which are now $+2v$ and $-2u$, is not important as this may be corrected by an adjustment to the gain of the colour signal channels. To take an electrical average of the chrominance signals belonging to adjacent lines, it is necessary to 'store' the information of one line for a period of one line so that it may be compared with the information of the next line. Thus, a delay line giving a delay of one line period (64 μs) is required in the receiver. All this may seem to be an unnecessary complication since in our example the hue is correct on both odd and even lines after the reinversion of the V component for the even lines. However, we have been considering the effect with no phase error in the received signal and the need for a delay line and electrical averaging will be seen when dealing with signals having phase errors.

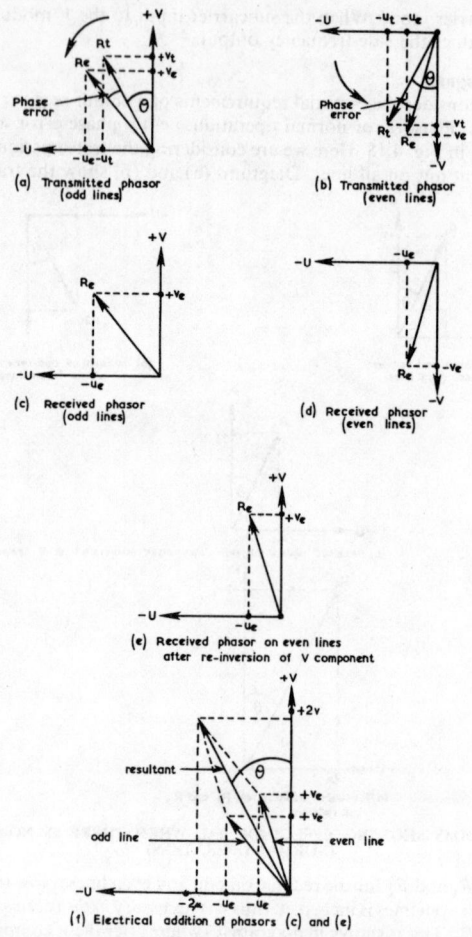

FIG. 4.19   DIAGRAMS SHOWING EFFECT OF PAL WHEN THERE IS A PHASE ERROR IN THE
RECEIVED SIGNAL

Fig. 4.19 shows the effect of the PAL modification when the received signal has a phase error. Suppose that a red hue R_t is transmitted on all lines but during transmission a phase advance occurs so that the actual signal transmitted is R_e. This is shown in diagrams (a) and (b) for odd and even lines where R_e is advanced on R_t by a small angle. Thus the phasors of the received chrominance signals on odd and even lines will be as shown in diagrams (c) and (d). In the receiver, the V component of the even line signal is reinverted, thus the even line phasor is now as in diagram (e). If the chrominance signals of (c) and (e) are now added together we get the resultant as shown in diagram (f). The even line signal has a phase which is lagging the true phase for red whilst the odd line signal has a phase which is advanced on the true phase for red. However, the phase of the electrical resultant is correct and a true red hue will be displayed. Note that the V and U components of the resultant are less than when there is no phase error [compare with Fig. 4.18(d)] thus the displayed signal will be slightly desaturated. Therefore, when a phase error occurs in the transmitting or receiving link, the effect of PAL is to cancel out the phase error resulting in the display of correct hues but with some small desaturation. Fortunately, the eye is not too sensitive to small desaturation errors.

At this point the basic operation of the heart of the PAL decoder in the receiver may be considered. Fig. 4.20 shows in schematic form the essential stages. The electrical averaging is carried out in the adder and subtractor blocks which are commonly

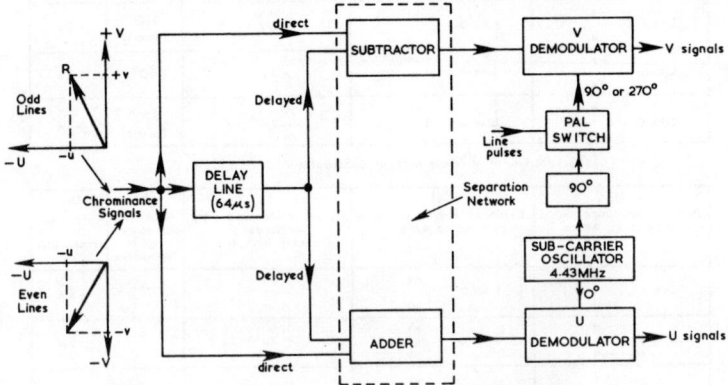

FIG. 4.20   THE HEART OF A PAL DECODER

referred to as the 'separation network'. The reason for using an adder and subtractor rather than just an adder is that in practice the V and U components of consecutive lines of chrominance signal are averaged separately. Both adder and subtractor have two inputs. One input is the DIRECT signal and the other is the DELAYED signal which is the output of the 64 μs delay line. It should be noted that because of the delay line, when, say, line 3 direct is arriving at the input to the adder or subtractor, line 2 delayed will be arriving at the input of these units.

Assume that a red hue is being transmitted on all lines. Thus the phasors on odd and even lines at the input to the delay line will be as shown. On odd lines the phasor has $+v$ and $-u$ components whereas on even lines the components are $-v$ and $-u$. Fig. 4.21 shows the result in column 3 of subtracting and adding the direct and delayed signals. On line 3 direct and line 2 delayed, for example, the subtractor gives out $+v - (-v) = +2v$ and $-u - (-u) = 0$, i.e. only the v component appears at the output of the subtractor. For these lines the adder gives out $+v + (-v) = 0$ and $-u + (-u) = -2u$, i.e. only the u component appears at the output of the adder. It will be noted that the output of the adder is $-2u$ on all lines but that the output of the subtractor is $+2v$ on one line and $-2v$ on the next line. The reason for this change in the subtractor output is because of the V component alternations at the transmitter.

Let us now turn to the demodulators. To demodulate the separated U and V components appearing at the outputs of the adder and subtractor, the original subcarrier must be made available. This is supplied from the subcarrier oscillator which has the same frequency and phase as the subcarrier fed to the balanced modulators. The subcarrier is fed directly to the U demodulator which demodulates the adder output to produce $-2u$ video signals on all lines. The subcarrier feed to the V demodulator is *via* a 90° phase advance network and the PAL switch. The PAL switch inverts the phase of the subcarrier input to the V demodulator on alternate lines, thus the phase will be 90° (with respect to the U demodulator phase) on one line and 270° on the next line. When the phase is 270°, the effect is to reinvert the V component of the signal supplied to the V demodulator. As a result, the V demodulator gives out $+2v$ video signals on all lines as is required in this case. The V and U outputs of the demodulators will, after amplification and processing in their respective channels, result in the correct red hue being displayed.

(1) Direct chrominance signal at input to subtractor		(2) Delayed chrominance signal at input to subtractor		(3) Output chrominance signal of subtractor (Direct − Delayed)	(4) Phase of sub-carrier at V demodulator	(5) Output of V demodulator
Line 3	+v −u	Line 2	−v −u	+ 2v	90°	+ 2v
Line 4	−v −u	Line 3	+v −u	− 2v	270° (invert)	+ 2v
Line 5	+v −u	Line 4	−v −u	+ 2v	90°	+ 2v
Line 6	−v −u	Line 5	+v −u	− 2v	270° (invert)	+ 2v

(a) Separation of V components in Subtractor

(1) Direct chrominance signal at input to Adder		(2) Delayed chrominance signal at input to Adder		(3) Output chrominance signal of Adder (Direct + Delayed)	(4) Phase of sub-carrier at U demodulator	(5) Output of U demodulator
Line 3	+v −u	Line 2	−v −u	− 2u	0°	− 2u
Line 4	−v −u	Line 3	+v −u	− 2u	0°	− 2u
Line 5	+v −u	Line 4	−v −u	− 2u	0°	− 2u
Line 6	−v −u	Line 5	+v −u	− 2u	0°	− 2u

(b) Separation of U components in Adder

FIG. 4.21   TABLES SHOWING RESULTS OF ELECTRICAL AVERAGING

Colour Bar Phasors

The effect of weighting and PAL may now be considered in relation to the phasors for the colour bar signals. Fig. 4.22 gives the values of the V and U components for fully saturated colour bars where

$$V = 0.877(E_R - E_Y) \text{ and } U = 0.493(E_B - E_Y).$$

The amplitude of the resultant phasors may be found from

$$\sqrt{V^2 + U^2}.$$

For example, on the yellow bar

$V = 0.097$ and $U = 0.44$ thus the resultant amplitude

$$= \sqrt{(0.097)^2 + (0.44)^2} = 0.45.$$

	W	Y	C	G	M	R	B	B
$E_R - E_Y$	0.00	+ 0.11	− 0.7	− 0.59	+ 0.59	+ 0.7	− 0.11	0.00
V	0.00	+0.097	−0.61	−0.52	+0.52	+0.61	−0.097	0.00
$E_B - E_Y$	0.00	− 0.89	+ 0.3	−0.59	+0.59	− 0.3	+0.89	0.00
U	0.00	−0.44	+0.15	−0.29	+0.29	−0.15	+0.44	0.00

FIG. 4.22   TABLE SHOWING V AND U VALUES ON FULLY SATURATED COLOUR BARS

The phase angle of yellow with respect to the $+U$ axis

$$= 180^\circ - \text{arc tan } \frac{V}{U} = 180^\circ - \text{arc tan } \frac{0.097}{0.44} = 167.6^\circ.$$

Fig. 4.23(a) shows the resultant amplitude and phases of the colour bar hues on odd lines. On even lines, due to the inversion of the V component, the transmitted colours have phases as shown in diagram (b). (Note the approximate phase positions as these can be useful in dealing with fault symptoms in the receiver decoder).

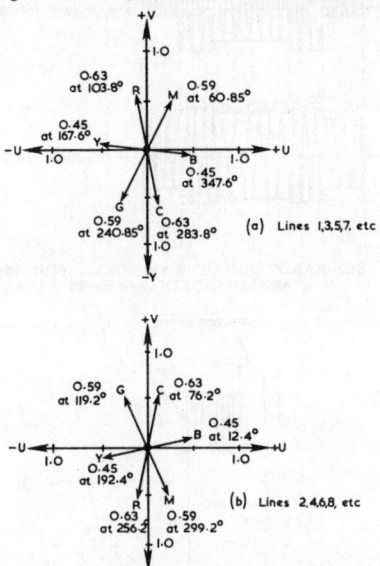

FIG. 4.23 PHASES OF COLOUR BAR SIGNALS ON ODD AND EVEN LINES (WEIGHTED VALUES)

Fig. 4.24 shows the waveforms for the U, V and combined sidebands when transmitting 95% saturated, 100% amplitude colour bars. These waveforms may be observed in the receiver decoder prior to synchronous demodulation of the V and U components (see Fig. 7.3).

THE COLOUR BURST SIGNAL

It was stated earlier that with suppressed-carrier operation the subcarrier must be made available at the receiver for successful detection of the colour information. Not only must the locally generated subcarrier be of the same frequency but it must also have the same phase as the subcarrier fed to the balanced modulators. Clearly, there must be some form of synchronisation between the transmitter and receiver subcarrier oscillators. This synchronisation is performed by the colour burst signal. The burst signal consists of approximately 10 cycles of the subcarrier transmitted on the back porch of the line sync. pulses, see Fig. 4.25. At the receiver these cycles are picked out by a gating circuit and then used to control the frequency and phase of the locally generated subcarrier.

The burst signal also conveys additional information to the receiver. This information is required to keep the PAL switch in the receiver decoder in step with the PAL switch at the transmitter. Without this information the receiver may reinvert the

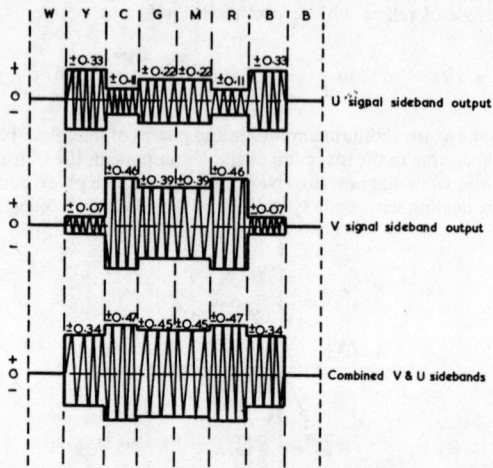

FIG. 4.24 V AND U SIDEBAND SIGNAL WAVEFORMS FOR 95% SATURATED, 100% AMPLITUDE COLOUR BARS

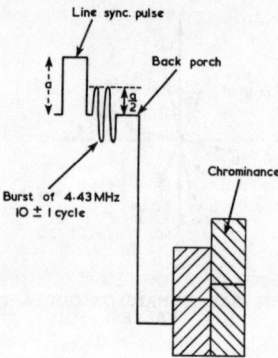

FIG. 4.25 THE COLOUR BURST

V component on the wrong lines resulting in the incorrect hue to be displayed. The additional information is conveyed by the burst phase, see Fig. 4.26. On odd lines the burst phase is 45° lagging on the − U axis whilst on even lines its phase is 45° leading on the − U axis. At the receiver this 'swinging burst' phase which is in sympathy with the V signal switching gives rise to an IDENTIFICATION SIGNAL which is used to synchronise the receiver PAL switch. It will be noted that the mean phase of the colour burst-lies along the − U axis. The reason for this particular mean phase rather than any other (+ U, +

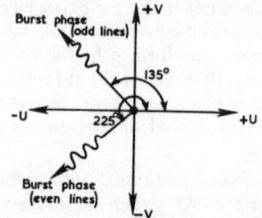

FIG. 4.26 THE SWINGING BURST

V or $- V$) is that the visibility of the burst, if not completely suppressed in the receiver, is minimal.

As there is no back porch on the field sync. pulses, it is not practicable to transmit a train of burst signals. Thus the burst is blanked out for a period of 9 lines each field during the equalising and field sync. pulse period. The burst blanking period is carefully chosen following a four-field sequence so that the phase of the burst when it reappears is always in the $+ V$ phase. This ensures that the locally generated subcarrier in the receiver is in the correct phase at the commencement of colour information at the start of each new field scan.

PAL ENCODER BLOCK SCHEMATIC

We may now summarise the important features of the PAL system of colour television discussed in this chapter with the aid of the block diagram given in Fig. 4.27.

The colour camera breaks down the coloured light emanating from the scene into the three basic components of red, green and blue light. Each camera tube produces at its output a primary signal voltage having an amplitude proportional to the amount of light falling on its face. These voltages are gamma corrected to give E'_R, E'_B and E'_G at the camera output. In block (1), portions of the primary signal voltages are added together to form the luminance signal E'_Y. In addition, the luminance signal voltage is subtracted from the E'_R and E'_B signal voltages to obtain the colour-difference signals $E'_R - E'_Y$ and $E'_B - E'_Y$. The colour-difference signals are then attenuated (weighted) in blocks 2 and 3 to produce the V and U signals where $V = 0.877 (E'_R - E'_Y)$ and $U = 0.493 (E'_B - E'_Y)$. The V and U signals are then fed to the balanced modulators via filters which limit the bandwidth to the required value of 0—1 MHz.

In blocks (6) and (7) the V and U signals amplitude modulate a common subcarrier of approximately 4·43 MHz, but with the subcarrier input to the V modulator in phase quadrature with the subcarrier input to the U modulator. The subcarrier is generated in block (10) and fed directly to the U modulator but via a 90° phase shift network (13) and the PAL switch (14) to the V modulator. The PAL switch is required to invert the phase of the subcarrier to the V modulator on alternate lines. Since balanced modulators are used the subcarrier is suppressed, thus the output of each modulator consists of sidebands only centred on 4·43 MHz. The chrominance sidebands are then combined in block (8) and subsequently fed to block (9) where they are combined with the luminance and sync. signals.

The line and field sync. pulses also blanking and equalising pulses are generated in block (11) with their timing controlled by frequency division of the subcarrier output of block (10). The sync. pulse generator also provides a half-line switching waveform for the PAL switch and line frequency pulses to the burst generator (12). The burst pulse generator supplies pulses of opposite polarity at line frequency having durations confined to within the back porch period to the V and U modulators. These pulses produce bursts of the subcarrier at the modulator outputs. On odd lines the individual bursts have phases of $+ V$ and $- U$ which, after combination in block (8), produce a burst at 135°. On even lines, due to the reversal in phase of the subcarrier input to the V modulator, the phases of the individual bursts are $- V$ and $- U$ which after combination provide a burst of 225° as is required.

Before the luminance signal is combined with the other signals on block (9) it is delayed for approximately 1 μs in block (15). This is necessary so that all signals arrive at the input of block (9) together, since the V and U signals are restricted in bandwidth compared to the luminance signal, i.e. they have longer rise-times than the luminance signal.

Fig. 4.28 shows the composite signal appearing at point X during one line of colour bars. A colour receiver requires all four components (a), (b), (c) and (d) to produce a well defined colour image on its screen. A monochrome receiver when tuned to the

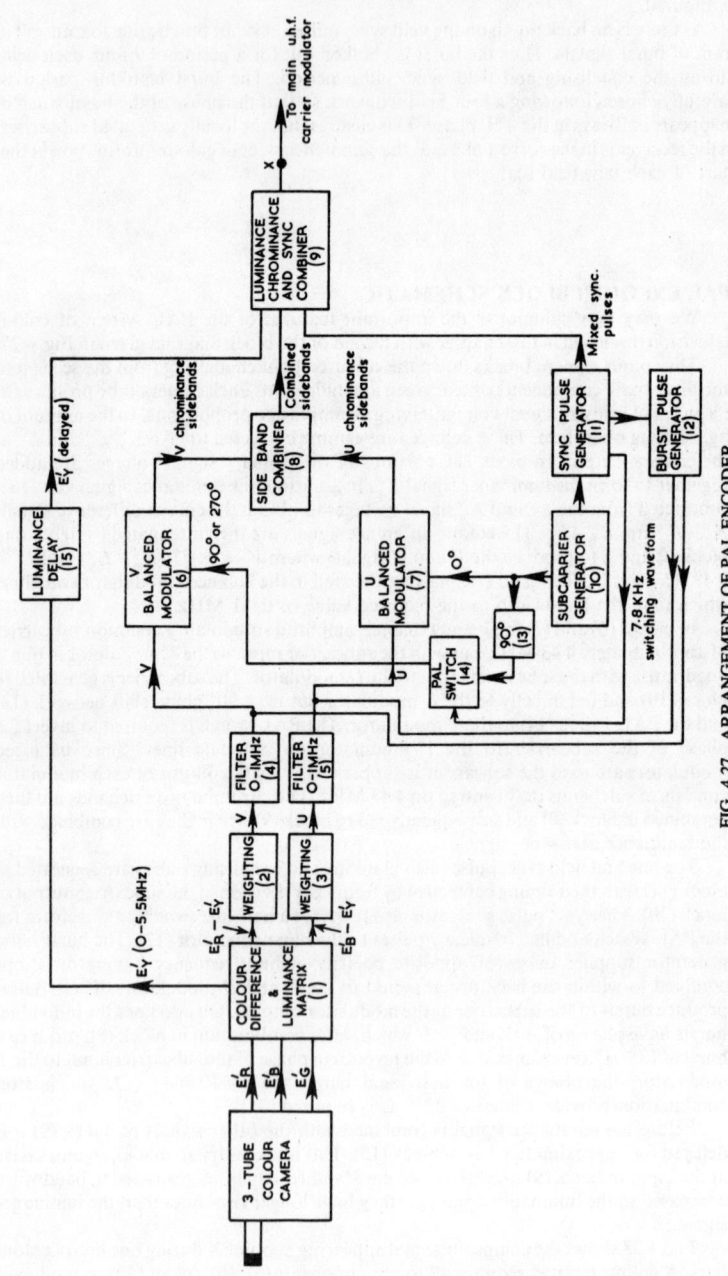

FIG. 4.27   ARRANGEMENT OF PAL ENCODER

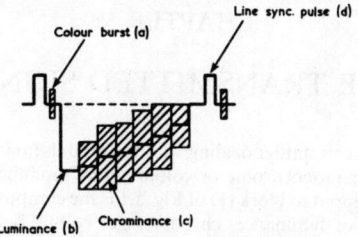

FIG. 4.28 COMPOSITE SIGNAL AT X (FIG. 4.27) DURING THE TRANSMISSION OF 95% SATURATED COLOUR BARS

same colour transmission requires only components (b) and (d). If the transmission is in monochrome then components (a) and (c) will not be transmitted.

CHAPTER 5

THE TRANSMITTED SIGNAL

THE section of the transmitter dealing with the modulation of the radiated carrier will accept either monochrome or colour signal modulation. During a colour programme, the input to block (1) of Fig. 5.1 is the composite colour modulation waveform consisting of luminance, chrominance, colour burst and sync. On the occasions when the programme is a black-and-white film or video recording, the input to block (1) is the luminance and sync. waveform only.

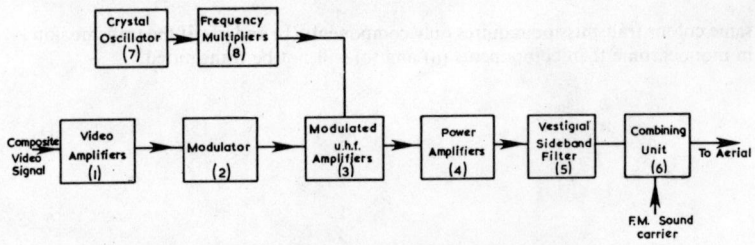

FIG. 5.1 BASIC TELEVISION TRANSMITTER

Amplitude modulation is used for the vision carrier which on 625-lines lies in the u.h.f. band. To obtain good frequency stability, the carrier is derived from a crystal oscillator working at a submultiple of the required frequency which is raised to the radiated frequency by the use of frequency multiplier stages. Amplitude modulation of the u.h.f. carrier takes place in block (3). The modulation levels which were originally set by the needs of monochrome operation are shown in Fig. 5.2. Maximum carrier, *i.e.* 100% modulation occurs on the tips of the sync. pulses whilst a minimum carrier of

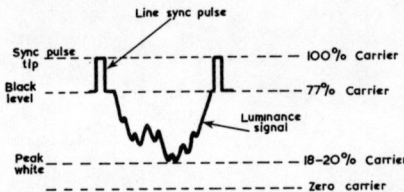

FIG. 5.2 MODULATION LEVELS FOR THE MONOCHROME (LUMINANCE) SIGNAL

18–20% modulation corresponds to peak white. Black and blanking are set at 77% modulation. This type of modulation is sometimes referred to as 'negative modulation' since increasing the amplitude of the video signal reduces the amplitude of the carrier. It may appear wasteful not to make use of the full range of modulation levels by limiting peak white at around 20% but there is a good reason for this. In the receiver, a technique called 'intercarrier sound' is used where an 'intercarrier sound i.f.' of 6 MHz is produced by beating the main vision and sound i.f.s together. If the vision carrier was returned to zero during modulation, there would be no intercarrier i.f. produced causing a loss of sound.

Fig. 5.3 shows the modulation levels on full amplitude, 95% saturated colour bars. In spite of the weighting given to the chrominance signals, the monochrome lower limit of 20% is exceeded for some hues. There is, however, some residual carrier present to meet the needs of intercarrier sound at the receiver. In a typical colour programme the lower level of 6% will not be reached since the colours will rarely be at full amplitude

58

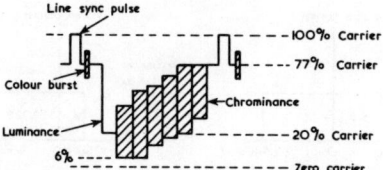

FIG. 5.3   MODULATION LEVELS ON 95% SATURATED COLOUR BARS

and highly saturated as for the colour bars which provide a critical test of the colour t.v. system.

The output of block (3) in Fig. 5.1 is the amplitude modulated u.h.f. carrier. Examples of the modulated carrier are shown in Figs. 5.4 and 5.5 for monochrome and colour operation. It should be noted that the chrominance modulation is a sine-wave type modulation (sidebands of 4·43 MHz) imposed upon the luminance levels.

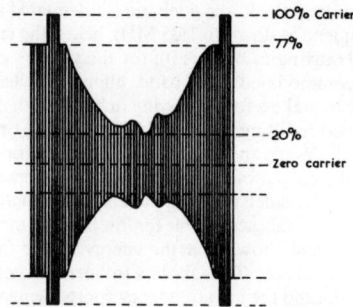

FIG. 5.4   TYPICAL MODULATED CARRIER WAVEFORM SHOWING SYNC. AND LUMINANCE
SIGNAL ENVELOPE

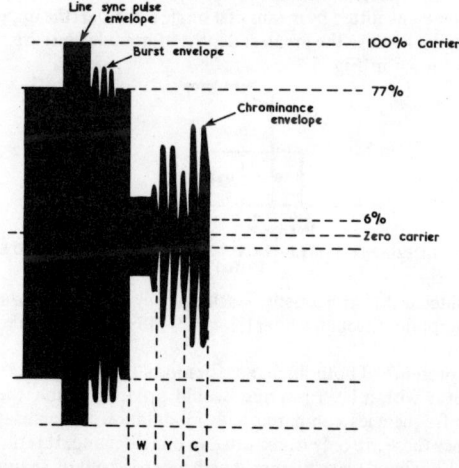

FIG. 5.5   MODULATED CARRIER WAVEFORM SHOWING PART COLOUR BAR MODULATION

VESTIGIAL SIDEBAND OPERATION

We have seen that the video bandwidth of the luminance signal extends up to 5·5 MHz. With amplitude modulation of the vision carrier we therefore get two sets of

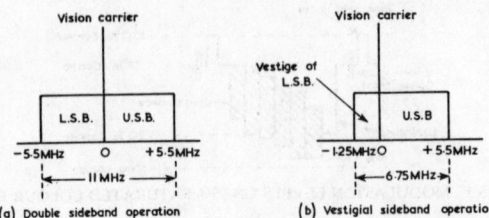

(a) Double sideband operation   (b) Vestigial sideband operation

FIG. 5.6 DIAGRAMS SHOWING THE BANDWIDTH SAVING BY EMPLOYING VESTIGIAL SIDEBAND WORKING

sidebands as shown in Fig. 5.6(a), producing an overall bandwidth of 11 MHz. Double sidebands transmission was used in the early days of television but is rather wasteful of bandwidth. Some of the side-frequencies of one sideband (upper or lower) may be removed leaving a vestige of that sideband and this idea is used in all modern systems. In the British 625-line system, the lower sideband is attenuated as in diagram (b) with only the lower side-frequencies down to 1·25 MHz below the carrier remaining. This results in a considerable saving in bandwidth for the vision signal, thereby allowing more channels to be accommodated in the bands allocated to television broadcasting.

It would be possible to make a further saving in bandwidth if one of the sidebands was completely suppressed resulting in single sideband (s.s.b.) operation and it may be asked why this is not done. When an ordinary envelope detector is used to demodulate an s.s.b. transmission, the output of the detector contains some distortion caused by the suppression of one of the sidebands. As the distortion increases rapidly with the depth of modulation, s.s.b. is only acceptable for low levels of modulation. An analysis of the television video signal shows that the energy of the frequency components reduces as the frequency increases. Thus the low frequency components will cause the highest modulation depths and the high frequency components will result in relatively low levels of modulation. Vestigial sideband operation is, in effect, a compromise between the bandwidth problem of double sideband working and the distortion problem of s.s.b. With vestigial sideband working, the high video frequencies (1·25 MHz—5·5 MHz) are transmitted by means of a single sideband (the upper) and the low video frequencies (which cause the greatest depths of modulation) are transmitted by both sidebands as shown in Fig. 5.7.

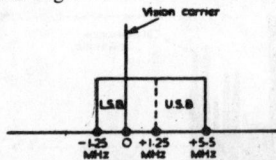

FIG. 5.7 VESTIGIAL SIDEBAND WORKING USING D.S.B. (0–1.25 MHz) AND S.S.B. (1.25 MHz–5 MHz)

The vestigial sideband characteristic is achieved by passing the output from the modulated u.h.f. amplifier through a filter [block (5) of Fig. 5.1] which removes part of the lower sideband.

Because of the presence of both sidebands for modulation frequencies from 0 Hz to 1·25 MHz, a receiver with a level response would produce double the output at its detector for these frequencies compared with modulation frequencies in the range 1·25—5·5 MHz since these are only transmitted in one sideband. It is thus necessary to correct for vestigial sideband transmission, which is carried out by shaping the receiver response as in Fig. 5.8 showing the ideal form. Clearly, the sum of symmetrical side-frequency components on either side of the vision carrier in the range 0—1·25 MHz will equal the amplitude of the side-frequency components from 1·25—5·5 MHz, e.g. $a+b=c$. In consequence, a receiver with this type of response will provide from its demodulator the same amplitude of output signal over the range of 0—5·5 MHz.

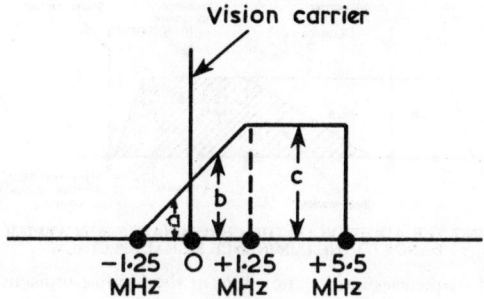

FIG. 5.8   RECEIVER RESPONSE REQUIRED TO CORRECT FOR VESTIGIAL OPERATION

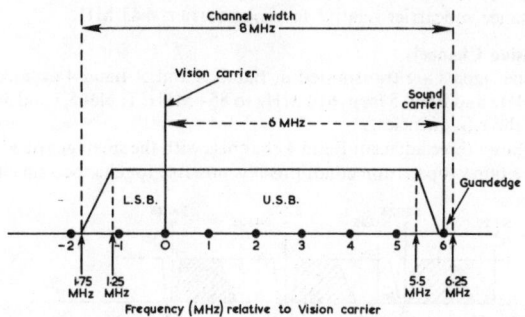

FIG. 5.9   625-LINE SPECTRUM OF TELEVISION SIGNAL (ONE CHANNEL)

Fig. 5.9 shows the bandwidth of the U.K. 625-line television signal including the sound carrier. The lower sideband of the vision signal is transmitted in full up to 1·25 MHz at which point attenuation is applied giving complete suppression at 1·75 MHz. In the upper sideband, frequencies up to 5·5 MHz are fully radiated but above this, attenuation is applied so that at 6 MHz suppression is complete. The sound carrier which is frequency modulated using a deviation of ± 50 kHz is fixed at 6 MHz above the vision carrier. The sound signal requires a bandwidth of about 180 kHz. This leaves a small guard edge to allow for inter-channel separation. By placing the sound carrier at the edge of the fully radiated sideband, the sound signal on demodulation at the receiver will fall just outside the bandpass of the video channel and so can be rejected from the video signal path. Because of the attenuated slopes on the vision sidebands, a full channel bandwidth of 8 MHz is required.

In Fig. 5.9 we have been considering the effect of luminance signal modulation only. During a colour transmission the chrominance signal which embraces 4·43 MHz ± 1 MHz in the video band will, upon modulation of the u.h.f. carrier, appear in the upper sideband of the transmitted vision signal as shown in Fig. 5.10. The chrominance signal has full sidebands extending up to 5·5 MHz and down to 3·1 MHz producing a full level bandwidth of approximately 2·3 MHz. The bandpass of the chrominance signal is slightly assymetrical which does not have a detrimental effect at the receiver. The chrominance signals are transmitted in one sideband only and it is no longer true that the higher modulating frequencies will result in small percentage modulations. As this can cause distortion of the detected chrominance signals, methods are used at the receiver to reduce the percentage modulation and hence the distortion. One method is to reduce the response of the vision i.f. amplifiers to the chrominance subcarrier, and details will be given later.

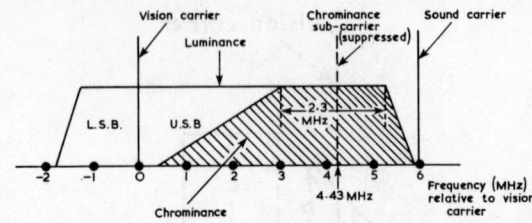

FIG. 5.10 SHOWING THE ADDITION OF THE CHROMINANCE SUBCARRIER AND ITS SIDE-BANDS TO THE LUMINANCE SIGNAL SPECTRUM

Some important frequencies and relationships of the 625-line transmission are:

> Channel bandwith: 8 MHz
> Sound carrier relative to vision carrier: 6 MHz
> Sound carrier relative to chrominance subcarrier: 1·57 MHz
> Chrominance subcarrier relative to vision carrier: 4·43 MHz

U.H.F. Television Channels

The 625-line signals are transmitted in Bands 4 and 5. Band 4 extends from 470 MHz to 582 MHz and Band 5 from 614 MHz to 854 MHz. Tables 5.1 and 5.2 show the channels and their frequencies.

Fig. 5.11 shows three adjacent Band 4 channels with the small guard edge between channels. In certain propagation condtions it is possible for channels on either side of

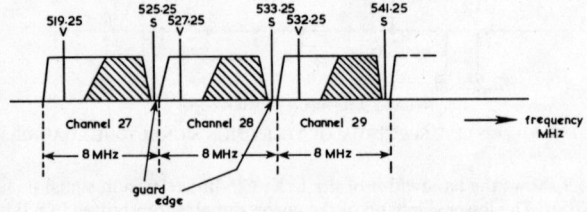

FIG. 5.11 THREE ADJACENT BAND 4 CHANNELS

the channel in use to cause interference. Protection against this type of interference is made by including adjacent channel rejector circuits in the receiver i.f. amplifier stages. If, for example, channel 28 is the channel in use, the sound carrier of channel 27 and the vision carrier of channel 29 may cause interference; these are called the 'adjacent channel sound' and the 'adjacent channel vision' respectively. These carriers being close to the channel in use produce intermediate frequencies in the receiver which lie close to the bandpass of the i.f. stages.

Local Station Channels

Each local area transmitter is allocated four channels. The lowest channel of a particular station is called n and the other channels are usually arranged according to the format of $n+3$, $n+6$, and $n+10$ or, alternatively, $n+4$, $n+7$ and $n+10$. Some examples of local-station channels are given under:

	BBC1	IBA 1	BBC 2	OTHER	AERIAL GROUP
CRYSTAL PALACE*	26	23	33	30	A red
EMLEY MOOR	44	47	51	41	B yellow
BELMONT	22	25	28	32	A red
WINTER HILL	55	59	62	65	C green
MENDIP	58	61	64	54	C green
WENVOE	44	41	51	47	B yellow

*Non-standard group.

	Channel	Frequency Range (MHz)	Vision Carrier (MHz)	Sound Carrier (MHz)
	21	470–478	471·25	477·25
	22	478–486	479·25	485·25
	23	486–494	487·25	493·25
	24	494–502	495·25	501·25
Aerial	25	502–510	503·25	509·25
Group A	26	510–518	511·25	517·25
(RED)	27	518–526	519·25	525·25
Channels	28	526–534	527·25	533·25
21–34	29	534–542	535·25	541·25
	30	542–550	543·25	549·25
	31	550–558	551·25	557·25
	32	558–566	559·25	565·25
	33	566–574	567·25	573·25
	34	574–582	575·25	581·25

TABLE 5.1 BAND 4 CHANNELS AND FREQUENCIES

Group	Channel	Frequency Range (MHz)	Vision Carrier (MHz)	Sound Carrier (MHz)
	39	614–622	615·25	621·25
	40	622–630	623·25	629·25
	41	630–638	631·25	637·25
Aerial	42	638–646	639·25	645·25
Group B	43	646–654	647·25	653·25
(YELLOW)	44	654–662	655·25	661·25
Channels	45	662–670	663·25	669·25
39–51	46	670–678	671·25	677·25
	47	678–686	679·25	685·25
	48	686–694	687·25	693·25
	49	694–702	695·25	701·25
Aerial Group E (BROWN) Channels 39–68	50	702–710	703·25	709·25
	51	710–718	711·25	717·25
	52	718–726	719·25	725·25
	53	726–734	727·25	733·25
	54	734–742	735·25	741·25
	55	742–750	743·25	749·25
Aerial Group D (BLUE) Channels 49–68 / Aerial Group C (GREEN) Channels 50–66	56	750–758	751·25	757·25
	57	758–766	759·25	765·25
	58	766–774	767·25	773·25
	59	774–782	775·25	781·25
	60	782–790	783·25	789·25
	61	790–798	791·25	797·35
	62	798–806	799–25	805·25
	63	806–814	807·25	813·25
	64	814–822	815·25	821·25
	65	822–830	823·25	829·25
	66	830–838	831·25	837·25
	67	838–846	839·25	845·25
	68	846–854	847·25	853·25

TABLE 5.2 BAND 5 CI'ANNELS AND FREQUENCIES

It will be noted some transmitters are allocated the same channel numbers, *e.g.* Emley Moor and Wenvoe. Mutual (co-channel) interference does not normally occur between such transmitters since they are far enough apart geographically. Including the lowest and highest channel in any group of a particular transmitting station there are 11 channel spacings. Thus, to receive the four local programmes (one not yet in use) efficiently, a receiving aerial should have a bandwidth of 11 × 8 MHz = 88 MHz.

CHAPTER 6

PROPAGATION, AERIALS AND FEEDERS

THE energy radiated from a television transmitter is in the form of plane electromagnetic waves. A 'plane wave' is one in which the electric and magnetic fields are mutually at right angles to one another and to the direction of propagation as in Fig. 6.1.

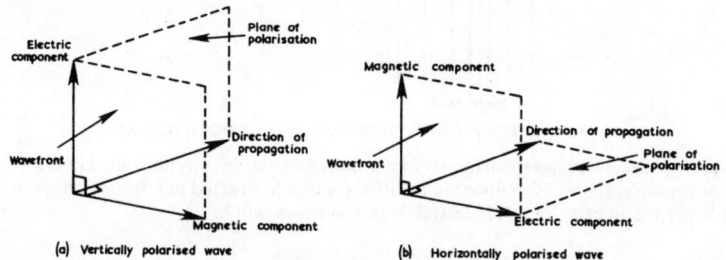

(a) Vertically polarised wave   (b) Horizontally polarised wave

FIG. 6.1 PLANE ELECTROMAGNETIC WAVES

The plane containing the electric and magnetic components of the wave is referred to as the 'wavefront'. To provide a reference, the plane containing the *electric field* and the direction of propagation is called the 'plane of polarisation'. Thus the wave of diagram (a) is said to be vertically polarised whilst the wave of diagram (b) is horizontally polarised. To intercept a wave and extract as much energy as possible from it, a receiving aerial must be mounted so that its conducting rods lie *parallel to the electric field*. Hence to receive a horizontally polarised wave the receiving aerial elements must be disposed horizontally.

A television wave is identical in character to a light wave since both are electromagnetic radiations. The television wave travels at the same speed of 3×10^8 metres per second and it obeys the same laws, *i.e.* it can be refracted and reflected. However, in one respect the television wave is different as its wavelength is longer than that of a light wave, but the shorter the wavelength of the radiated television signal the greater is the similarity to light radiations.

FIELD STRENGTH

The field strength of a wave radiated from a transmitter increases as the frequency of the transmitter aerial current increases, but is inversely proportional to the distance from the transmitter. In view of the former one would expect a greater received signal from a Band 5 transmission than, say, a Band 3 transmission at the same distance from the transmitter, but this is not the case. A transmitter radiates a certain number of kilowatts of energy which is dissipated in the form of a 'field' that becomes weaker as it progresses from the transmitter. The 'field' can be expressed as the p.d. set up between any two points lying parallel to the electric component and spaced one metre apart. Field strength is usually quoted as so many 'volts per metre' or in practice mV per metre or μV per metre. Practically all television receiving aerials are based on the half-wave dipole. This type of aerial has a physical length of approximately $\lambda/2$ where λ is the wavelength of the transmission. Thus the higher frequency of the transmission, the shorter the length of the dipole aerial and hence the smaller the 'field' that it intercepts.

The open-circuit e.m.f. (E) of a half-wave dipole is given by

$$E = \frac{\lambda}{\pi}e \qquad (6.1)$$

where λ is the wavelength in metres and e is the field strength in volts per metre.

65

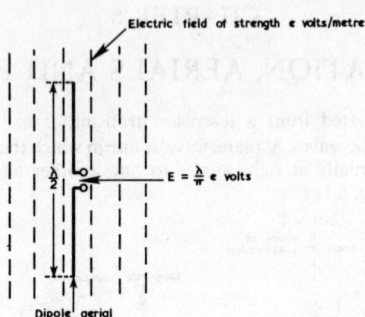

FIG. 6.2   E.M.F. INDUCED IN HALF-WAVE DIPOLE AERIAL

This expression applies when the aerial is disposed to the 'field' as illustrated in Fig. 6.2. For example, the e.m.f. induced in a half-wave dipole situated in a field strength of 5 mV per metre of a 200 MHz (Band 3) transmission will be

$$E = \frac{1 \cdot 5 \times 5 \times 10^{-3}}{\pi} \text{ volts} \qquad \text{(200 MHz = 1·5 metres wavelength)}$$

$$\simeq 2 \cdot 4 \text{ mV}.$$

With identical field strength but with transmission at a frequency of 800 MHz (Band 5), the e.m.f. would be

$$E = \frac{0 \cdot 375 \times 5 \times 10^{-3}}{\pi} \text{ volts} \qquad \text{(800 MHz = 0·375 metre wavelength)}$$

$$\simeq 0 \cdot 6 \text{ mV}.$$

Thus for similar *radiated* powers, the e.m.f. induced in a plain dipole operating at 800 MHz is a quarter of the e.m.f. induced in a plain dipole operating at 200 MHz. To overcome the reduced signal pick-up at u.h.f. as compared with v.h.f., the power radiated from the transmitter may be increased. However, to radiate very high powers is expensive and it is more economical to obtain the required increase in signal pick-up at u.h.f. partly by increasing the transmitter power and by increasing the transmitting aerial height and gain.

The power output of a television transmitter is normally quoted as 'so many kilowatts' e.r.p. (effective radiated power). Effective radiated power is a parameter which combines transmitter power with *aerial gain*. By using a directive aerial system the transmitter energy may be beamed in a particular direction thereby increasing the signal pick-up in that direction. This, of course, is at the expense of a reduction in field strength in other directions and arises out of the 'aerial gain.' It is usual to express aerial gain as a ratio of the powers required at the input of the aerial under consideration and a plain dipole in order to produce the same field strength in a given direction.

SPACE WAVE PROPAGATION

The energy radiated from a transmitting aerial may be directed so as to take any of three paths. The wave may be directed upwards towards the ionosphere (sky wave), exist close to the ground and follow the earth's curvature (ground wave) or travel in the space immediately above the earth (space wave). It is the space wave we are interested in as it is principally used by systems operating at high frequencies, *e.g.* television, f.m. radio and radar.

The range of the space wave is chiefly limited by the height of the transmitting and receiving aerials. As Fig. 6.3 shows, a receiving aerial R_1 will lie in the line-of-sight path of the transmitting aerial T and in consequence will receive its radiations. A receiving

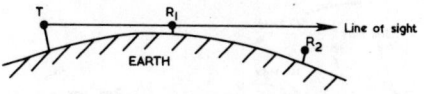

FIG. 6.3 SPACE WAVE RANGE LIMITED TO LINE OF SIGHT

aerial R_2 at the same height as R_1 but located farther round the earth's circumference cannot 'see' the transmitting aerial and therefore does not pick-up the line-of-sight signal.

At v.h.f. and u.h.f., reflections from the ground or some other object becomes more important. Consider Fig. 6.4 which shows a wave P_2 reflected from the ground arriving at the aerial R in addition to the direct wave P_1. This may seem to be advantageous but

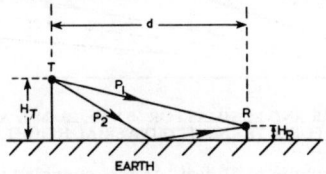

FIG. 6.4 DIRECT WAVE P_1 AND GROUND WAVE P_2

when P_2 is reflected from the ground the phase of the wave changes by 180°; thus P_1 and P_2 tend to cancel at the receiving aerial. If the path lengths of the two signals were identical, P_1 and P_2 would exactly cancel and no signal voltage would arrive at the receiver. In these circumstances the only reason why a signal is fed to the receiver is that the path of P_2 will always be longer than that of P_1. The field strength of the wave arriving at R in Fig. 6.4 may be found from

$$E = \frac{68 \cdot 8 \sqrt{P} \, H_T H_R}{d^2 \lambda} \text{ volts per metre} \tag{6.2}$$

where P = transmitter power (watts), H_T = transmitting aerial height (metres), H_R = receiving aerial height (metres), d = distance between transmitting and receiving aerials (metres) and λ = wavelength in metres.

The expression shows that the receiving aerial height is an important factor. By raising the height, the signal arriving at the receiver may be increased. As the aerial height is increased, the path length of P_2 is increased which increases the phase difference between P_1 and P_2, thus providing a larger signal. If the aerial height is increased until the difference in path lengths corresponds to $\lambda/2$, the signal arriving at the receiver will rise to a maximum. Raising the aerial height beyond this point will result in a fall in signal strength as the difference in path lengths approaches one wavelength. When the difference in path lengths is exactly one wavelength, the signal will be zero. A further increase in aerial height will cause the signal arriving at the receiver to increase and when the difference in the path lengths is exactly $3\lambda/2$ the signal will again be at a maximum, and so on.

Fig. 6.5 shows the result of calculations based on this effect assuming a transmitting aerial height of 1000 feet, which is fairly typical. At 50 MHz (Band 1) the first signal maximum occurs at about 130 feet at a range of 5 miles. As the chimney-height of a two-storey building is about 30 feet, the first signal maximum is clearly out of reach for most domestic aerial installations. Thus the rule for *Band 1* is to *mount the aerial as high as possible aiming for the first signal maximum*. At 200 MHz (Band 3) the first signal maximum occurs at about 32 feet at 5 miles with the first signal minimum occurring at about 48 feet. Thus, at close ranges on *Band 3 it is worthwhile moving the aerial up or down* in case the aerial lies in an area of minimum signal. At 800 MHz (Band 5), the first signal maximum occurs at about 8 feet at 5 miles falling to a minimum at 12 feet but

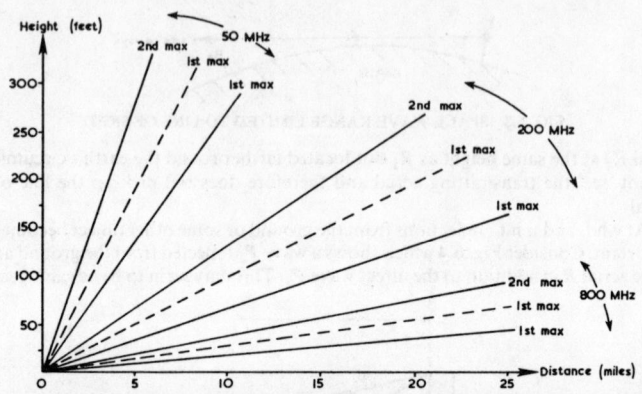

FIG. 6.5   FIELD MAXIMAS AND MINIMAS FOR 50 MHz, 200 MHz AND 800 MHz CALCULATED
FOR A TRANSMITTED AERIAL HEIGHT OF 1000 FEET

rising to a second maximum at 16 feet. As it is possible to pass from maximum to
minimum signal levels within the space of a few feet, the rule for *Bands 4 and 5 is again
to move the aerial up or down* from its initial setting to see if the signal strength may be
improved. The results set out in Fig. 6.5 are based on transmissions over a 'flat' surface
and should not be relied on. They do however give an indication as to what may be
happening at close and medium ranges of a transmitter. As regards field strength there
appears to be virtually no difference between vertically and horizontally polarised
waves and both are used for television transmissions.

It will be noted from expression (**6.2**) that the field strength increases with
frequency. Any benefit gained from this is exactly cancelled by the fact that the voltage
delivered to an aerial varies inversely with frequency; see expression (**6.1**).

EFFECT OF EARTH'S CURVATURE

Television reception is limited to the line-of-sight path of the transmitted wave.
With a transmitting aerial height of 1000 feet, the distance to the horizon is about 38
miles, see Fig. 6.6. Obviously, a high elevation is desirable for the transmitting aerial as
it will see a more distant horizon thereby increasing the service area. It is important that

FIG. 6.6   HORIZON OF 38 MILES FROM TRANSMITTING AERIAL HEIGHT OF 1000 FEET

the maximum power radiated from the aerial should not be horizontal to the earth as it
will never reach the earth at all. The transmitting aerial is designed therefore to 'tilt' the
beam downwards by half or three-quarters of a degree so that the maximum power is
directed towards the edges of the service area.

The area served by a t.v. transmitter may be greater than the distance to the
horizon, due mainly to the phenomenon of 'diffraction'. When a television wave strikes
the terrain at the horizon it is diffracted, *i.e.* it spreads downwards. The idea is shown in
Fig. 6.7 which compares the effect for Bands 1, 3 and 5 transmissions. The lower the
frequency of the wave, the more the diffraction. At Band 5 frequencies there is little

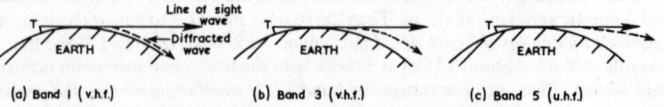

FIG. 6.7   DIFFRACTION OF THE WAVE AT THE HORIZON INCREASING THE SERVICE
AREA OF A TRANSMITTER

diffraction as u.h.f. waves behave more like light waves. Thus, beyond the horizon it is possible to pick up the television wave, but the field strength falls off rapidly particularly so at u.h.f. where the area beyond the horizon is in deep shadow.

Another important factor affecting the propagation of the television signal is the terrain of the country in the path of the wave, especially at high frequencies. Hills, large buildings and forests, etc. can create local shadow zones as illustrated in Fig. 6.8. A receiving aerial lying deep in the shadow zone will be starved of signal. It is this effect

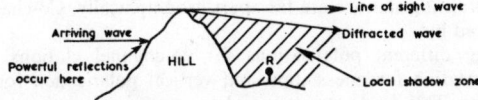

FIG. 6.8   LOCAL SHADOW ZONE CREATED BY LARGE HILL

more than diffraction which reduces the service area of Bands 4 and 5 transmissions as compared with Bands 1 and 3 which have roughly the same service areas. Trees are a serious obstacle to u.h.f. propagation, particularly when in full leaf. It appears that the branches introduce a good deal more loss of signal strength than the trunks, so if it is impracticable to raise the aerial above the trees it might be beneficial to lower it, provided that this did not encounter any other obstruction.

EFFECT OF THE TROPOSPHERE

Reception in the fringe area of a transmitter is affected by refraction of high angle waves from the transmitter by the troposphere, see Fig. 6.9. The troposphere is the

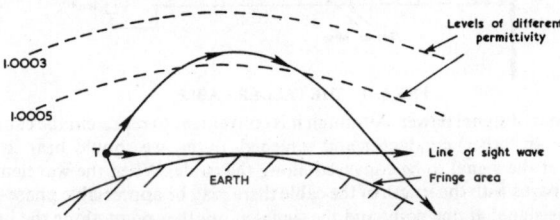

FIG. 6.9   REFRACTION BY THE TROPOSPHERE

region of changing weather and clouds extending from sea level up to an altitude of about 10 miles. Refraction of television waves by this part of the earth's atmosphere is affected by the pressure, temperature and water vapour content of the region. Because of water vapour in particular, the permittivity of the air in this region is greater than unity. In general the permittivity decreases with increase in height. The decrease in permittivity and, hence refractive index, causes high angle waves to be bent back towards the earth. The television wave thus travels some distance around the earth's curvature into the shadow zone of the line-of-sight signal.

It may be thought that the effect of tropospheric propagation would be beneficial as regards fringe area reception. However, signals refracted by the troposphere vary in amplitude with changes in conditions of the troposphere and causing fading of the received picture. As the strength of the signal arriving from the troposphere may exceed the 'direct' signal available in the fringe area, severe fading occurs during certain weather conditions and reception is unreliable. Thus the service area of a transmitter is limited by signals received from the troposphere.

Signals propagated via the troposphere are important for another reason. During stable weather conditions, following a hot day a 'duct' is formed between layers in the troposphere which 'traps' the signal and causes many reflections as indicated in Fig. 6.10. In such conditions the signal can be propagated over distances up to several hundred miles. These 'ducted' signals can cause interference between stations operating on common channels (co-channel interference), in spite of the fact that co-

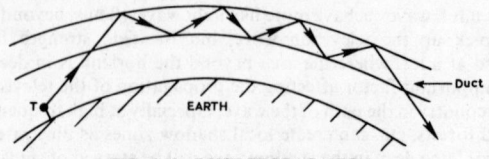

FIG. 6.10 DUCT PROPAGATION

channel transmitting stations are far apart geographically. Co-channel interference
may be reduced by:

(a)   Using different polarisations for co-channel ctations, *e.g.* horizontal
       polarisation for one station and vertical polarisation for the co-channel
       station. This does not give complete protection because propagation effects
       causes 'cross polarisation'.

(b)   'Off-setting' the transmitter frequencies by one-third or two-thirds of line
       frequency. This has the effect of reducing the visibility of the interfering
       pattern.

THE AERIAL FEEDER CABLE

The feeder cable connected between aerial and receiver, see Fig. 6.11, is employed
to convey the received signal energy from aerial to receiver and should do so with

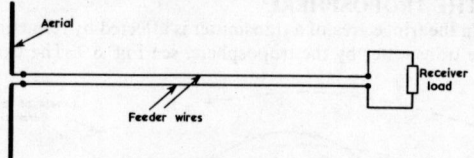

FIG. 6.11   THE FEEDER CABLE

negligible loss of signal power. Although it is convenient to represent the cable as a pair
of wires or an 'inner' conductor and screened 'outer' we should bear in mind the
frequency of the signal to be conveyed along the cable. When the wavelength of the
signal compares with the length of the cable there may be appreciable phase-difference
between the signal at one point and the signal at another point along the line. This is
because a pair of feeder wires possess distributed inductance and capacitance which is
usually ignored in the distribution of low frequency signals.

Fig. 6.12 shows an approximate equivalent circuit for a pair of feeder wires. The
quantities L, C, R_1 and R_2 are evenly distributed as an infinite number of infinitely
small 'lumps' along the length of the feeder of Fig. 6.11. Suppose a signal voltage is
applied to the input of the first section. The voltage will cause a current to flow in the
first capacitor but the current will lag the input voltage because of the inductance of the
wires. As the capacitor charges, a voltage will build up across it but delayed on the
input voltage. This voltage in turn will feed current into the next section causing the
second capacitor to charge up and the voltage across it will be further delayed on the
input voltage. In its turn the voltage across the second capacitor drives current into the
next section, and so on. In this way the signal is transmitted along the cable until it

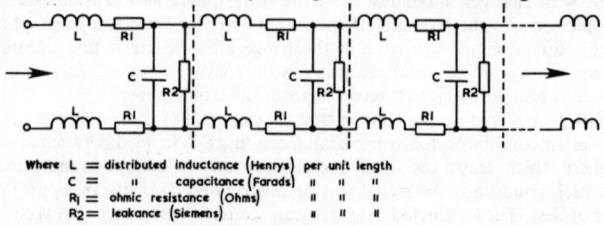

Where L = distributed inductance (Henrys) per unit length
       C =   "   capacitance (Farads)   "   "   "
       R_1 = ohmic resistance (Ohms)   "   "   "
       R_2 = leakance (Siemens)   "   "   "

FIG. 6.12   ELECTRICAL REPRESENTATION OF FEEDER CABLE

reaches the load at the far end. The amount of phase lag depends upon the distributed inductance and capacitance per unit length and is related to them by:

Phase change coefficient $\beta = \omega \sqrt{L.C}$ measured in radians per unit length (when the line has no losses).

The length of line at which the current or voltage for the first time is in phase with the supply end is called the 'wavelength'. The feeder wavelength will be somewhat shorter than the 'free space' wavelength since the velocity of the wave in the feeder is less than in free space. The velocity of the wave may be found from:

$$v = \frac{3 \times 10^8}{\sqrt{\mu_r e_r}} \text{ metres/sec.}$$

where μ_r and e_r are the relative permeability and permittivity of the cable dielectric.

Characteristic Resistance (R_o)

If we take an infinite length of feeder (with negligible losses) as in Fig. 6.13 and apply a signal voltage V, a signal current I will flow in the line (the distributed

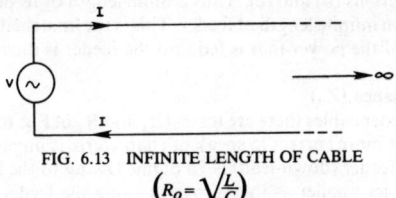

FIG. 6.13   INFINITE LENGTH OF CABLE

$$\left(R_o = \sqrt{\frac{L}{C}} \right)$$

components of the line providing the current path). The ratio V/I r.m.s. is called the 'characteristic resistance' of the feeder. If losses are neglected, R_o may be found from

$$R_o = \sqrt{\frac{L}{C}}$$

Note that since we are considering an infinite length of feeder, the wave does not reach the end and thus does not 'know' what the load is (if any). Therefore, the load value does not affect the value of R_o which is entirely dependent upon the parameters of the cable itself. Note also that frequency does not come into the expression, *i.e.* R_o is the same for all frequencies.

As an example suppose that a cable has a distributed inductance of 1 mH per mile and a distributed capacitance of 0.25 μF per mile, then

$$R_o = \sqrt{\frac{1 \times 10^{-3}}{0.25 \times 10^{-6}}} \text{ ohms} = 63 \text{ ohms}$$

If we approximate R_o to 60 ohms, then if a signal voltage of 12 m V is applied at the input of the feeder with this characteristic resistance, the current in the cable wires will be

$$\frac{12 \times 10^{-3}}{60} = 0.2 \text{ mA (see Fig.6.14).}$$

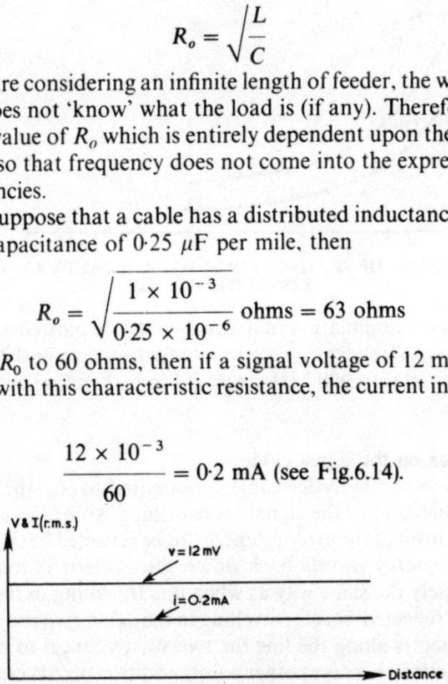

V & I(r.m.s.)

v = 12 mV

i = 0·2mA

Distance

FIG. 6.14   VOLTAGE AND CURRENT IN AN INFINITE LENGTH OF CABLE OF $R_o = 60 \ \Omega$ (NO LOSSES)

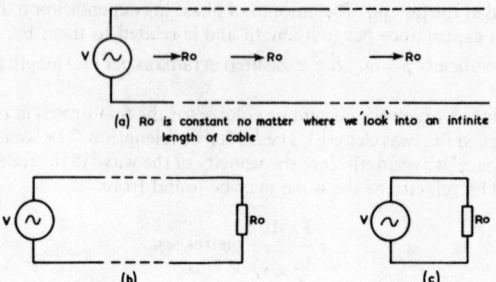

FIG. 6.15 A FINITE LENGTH OF CABLE TERMINATED IN R_0 'LOOKS' LIKE AN INFINITE LENGTH

Since R_0 is constant no matter where we 'look' into an infinite length of feeder, [see Fig. 6.15(a)], any of the sections may be removed and replaced by a resistor of value equal to R_0 as in diagrams (b) and (c). Thus a finite length of feeder terminated in R_0 behaves the same as an infinite length of feeder. This is the 'matched' or 'reflection-free' condition where all of the power that is fed into the feeder is dissipated in the load.

Characteristic Impedance (Z_0)

In all practical feeder cables there are losses (R_1 and R_2 of Fig. 6.12) which make the feeder reactive so it is more correct to speak of characteristic impedance. A typical Z_0 of a receiving aerial feeder (down-lead) is 75 ohms. Owing to the losses in the feeder, voltage and current get smaller as they progress along the feeder, see Fig. 6.16. The attenuation follows a logarithmic law and it is usual to express the attenuation in decibels per unit length (normally per 100 feet or per 10 m).

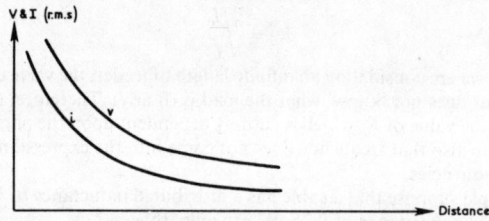

FIG. 6.16 ATTENUATION OF V AND I ALONG A PRACTICAL CABLE FOLLOWS A LOGARITHMIC LAW

Fig. 6.17 shows the attenuation of a 'low-loss' cable plotted against feeder length. The losses increase with frequency, e.g. the attenuation may be 0·75 dB per 10 m at 100 MHz but 2·6 dB per 10 m at 900 MHz. At u.h.f. it is therefore best to use the shortest feeder run possible.

Effect of a mismatch on the signal cable

If the termination of the feeder cable is not equal to Z_0, the feeder is said to be mismatched and reflection of the signal wave occurs. An o/c or s/c termination as in Fig. 6.18 will result in all of the arriving energy to be reflected back towards the sending end. As the signal energy travels back down the cable it is retarded in phase and attenuated in precisely the same way as when it is travelling in the forward direction. The 'forward' and 'reflected' waves travelling in the cable give rise to a 'standing-wave' pattern. At some points along the line the two waves cancel to produce voltage and current minima (nodes) whereas at other points addition occurs creating maxima (anti-nodes). The maxima and minima repeat themselves regularly at half-wavelength intervals. In diagram (a) the voltage at the termination must be zero since it is a s/c

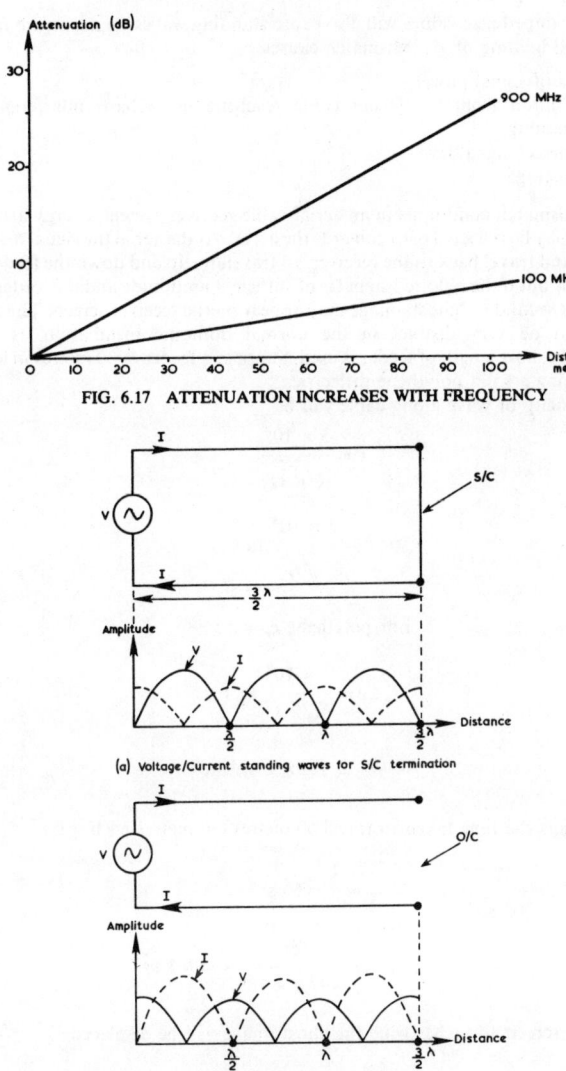

FIG. 6.17 ATTENUATION INCREASES WITH FREQUENCY

(a) Voltage/Current standing waves for S/C termination

(b) Voltage/Current standing waves for O/C termination

FIG. 6.18 EXTREME EXAMPLES OF MISMATCH

whereas in diagram (b) the current is zero at the termination as it is o/c. Both situations represent a gross mismatch of the feeder cable.

The reason that reflection occurs at the termination when it is mismatched to the cable is that the wave does not 'know' what the termination is until it arrives there. A parallel situation occurs when a sound wave is directed at a wall. If the wall is made of an acoustically absorbent material the sound energy will be absorbed; this is the matched condition. If the wall is made of brick, reflection will occur and the sound wave will travel back to the source.

Of course, the s/c and o/c terminations are extreme examples of mismatch.

Intermediate impedance values will also cause standing waves on departure from the nominal ideal loading of Z_0. Mismatch causes:

(a)   Loss of signal power.
(b)   Radiation from the feeder cable resulting in perhaps interference and patterning.
(c)   Incorrect signal levels.
(d)   Ghosting.

Under mismatch conditions in an aerial-cable-receiver system, energy arriving at the receiver may be reflected back towards the aerial. At the aerial the signal may again be reflected and travel back to the receiver. In travelling up and down the feeder some loss will occur but if the reflected signal is of sufficient amplitude and the feeder length appreciable a second or 'ghost' image may appear on the receiver screen. The ghost is not likely to be very distinct in the normal domestic installation as only a comparatively short length of feeder is used. Suppose a feeder of 30 metres in length is employed using a solid polythene dielectric.

Now velocity of wave along cable will be

$$v = \frac{3 \times 10^8}{\sqrt{\mu_r e_r}}$$

$$\text{or } v \simeq \frac{3 \times 10^8}{\sqrt{e_r}} \text{ m/s}$$

For polythene $e_r = 2 \cdot 3$

$$\therefore v = \frac{3 \times 10^8}{\sqrt{2 \cdot 3}} \text{ m/s}$$

$$= 197 \cdot 8 \times 10^6 \text{ m/s}$$

Thus the time taken to travel 60 metres (30 metres each way)

$$= \frac{1}{v} \times 60 \text{ s}$$

$$= \frac{60}{197 \cdot 8 \times 10^6} \text{ s} \simeq 0 \cdot 3 \ \mu s.$$

On a t.v. screen 14 inches wide the ghost image will be displaced

$$\frac{0 \cdot 3}{52} \times 14 \quad \text{inches}$$

(note active line period is $52\mu s$) to the right of the main image, which is approximately $\frac{1}{25}$ inch. Although the ghost may not be visible as a distinct second image the definition of the picture may be impaired.

Coaxial Cable

The most common type of television down-lead is coaxial cable and the usual form of construction is as in Fig. 6.19(a). The cable consists of a solid 'inner' copper conductor (or twisted wires) surrounded by a flexible insulating material such as polythene. The insulating medium may be of solid polythene or of cellular construction

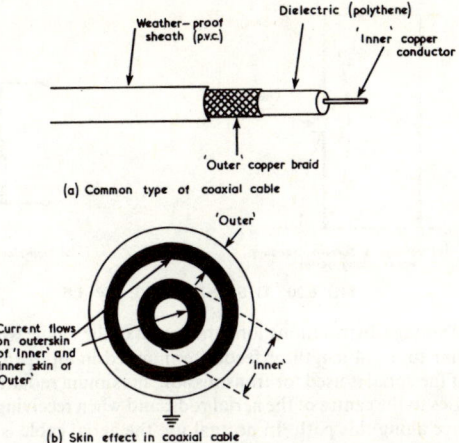

(a) Common type of coaxial cable

(b) Skin effect in coaxial cable

FIG. 6.19 COAXIAL CABLE

to reduce the losses at high frequencies. The 'outer' conductor consists of a copper braid wound over the dielectric. To protect the cable from moisture and damage it is covered with a tough p.v.c. sheath.

At high frequencies, the current in a conductor tends to flow in its outer skin ('skin effect'). As the frequency is raised, the skin becomes thinner and the cross-sectional area smaller. Since the resistivity of a conductor is inversely proportional to the cross-sectional area, the a.c. resistance of a cable increases with frequency (proportional to $\sqrt{f}$). With a coaxial cable the skin is formed on the outside of the 'inner' conductor but on the inside of the 'outer', see diagram (b). The penetration of current on the 'outer' braiding is negligible (except under high-standing wave conditions) thus the screen may be earthed. The earthing of the screen helps to suppress the general pick up of unwanted signals. The screen is normally connected to the receiver chassis *via* a small value capacitor which places the screen at neutral mains potential as regards signals. The coaxial feeder is an unbalanced cable, *i.e.* its wires are unbalanced to earth.

As it is important to maintain correct matching in an aerial installation and continuous screening of the feeder cable, lengths of cable must not be joined by way of a junction box or barrier strip as for ordinary electrical wiring. A properly designed 'line-coupler' should be used which maintains the Z_o at the joint and provides continuation of the screening. Feeders of different characteristic impedances should not be joined together and all plugs and sockets should be used only with the cable Z_o they were designed for.

Coaxial cable of 75Ω characteristic impedance is normally used as this impedance value provides a match to the centre impedance of a half-way dipole aerial.

RECEIVING AERIALS

Aerials intended for television signal reception at v.h.f. and u.h.f. are invariably based on the half-way dipole, see Fig. 6.20. The aerial consists of two rods with an overall length of

$$\frac{\lambda}{2} \text{ (where } \lambda = \frac{v}{f} \text{) less about 5\%.}$$

At 900 MHz the half-wave dipole would have a length of about

$$\frac{3 \times 10^8}{900 \times 10^6 \times 2} \times \frac{95}{100} \text{ metres} \simeq 15 \cdot 8 \text{ cm.}$$

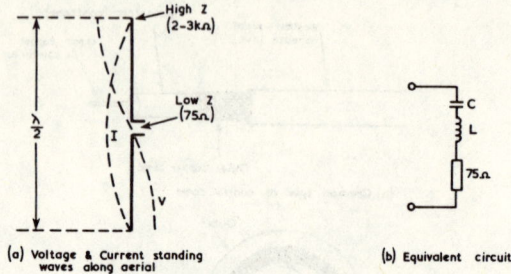

(a) Voltage & Current standing
     waves along aerial

(b) Equivalent circuit

FIG. 6.20   THE HALF-WAVE DIPOLE

The current and voltage distribution along the length of the dipole when energised with
a signal is similar to a $\lambda/4$ length of feeder terminated in an open circuit as in Fig.
6.18(b). Thus, if the aerial is used for transmission, maximum radiation occurs along a
line at right angles to the centre of the aerial rods; and when receiving, maximum pick-
up will take place along this path. In normal use the aerial cable is connected to the
centre where the impedance is low at about 75 ohms. When the dipole has a length so
that it corresponds to approximately $\lambda/2$ of the arriving signal, the aerial behaves as a
resonant circuit, see diagram (b). The aerial thus responds more to certain frequencies
than to others just like a tuned circuit.

Polar Diagrams

The directional performance of an aerial may be indicated by a polar diagram in
which the distance from the origin to the curve represents the relative signal strength of
the received or radiated signal in that direction.

For a vertically mounted dipole, the polar diagrams are as in Fig. 6.21. Diagram (a)
shows a plan view looking down on the end of one of the aerial rods. In this case the

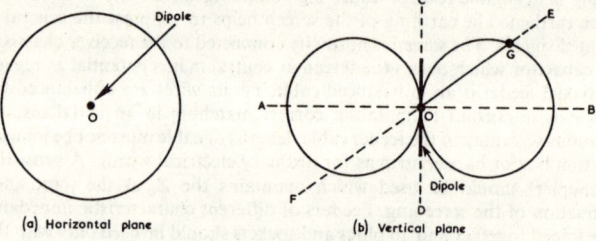

(a) Horizontal plane

(b) Vertical plane

FIG. 6.21   POLAR DIAGRAMS OF VERTICAL DIPOLE AERIAL

distance from the origin O to the curve (a circle) is the same no matter what direction is
considered. Thus the aerial is completely non-directional or 'omni-directional' in the
horizontal plane. This is not so in the vertical plane, diagram (b). Here the aerial has
maximum response to waves arriving along the path $A–B$ but zero response to waves
arriving along the path $C–D$. A wave arriving from E along a path $E–F$ would have a
relative strength as indicated by the solid line OG.

For a dipole mounted to receive horizontally polarised waves, Fig. 6.21(a) is the
polar diagram in the vertical plane and (b) in the horizontal plane. It will thus be seen
that a vertically mounted dipole offers no rejection to unwanted signals arriving at
small angles to the aerial. The horizontally mounted plain dipole, however, offers
rejection to an unwanted signal if the aerial is so orientated that the path of the
unwanted signal corresponds to the direction $C–D$. In its basic form the plain dipole is
only suitable in areas of high signal strength where reflections are of little importance.
For these reasons, the plain dipole is normally used for receiving Band 1 transmissions

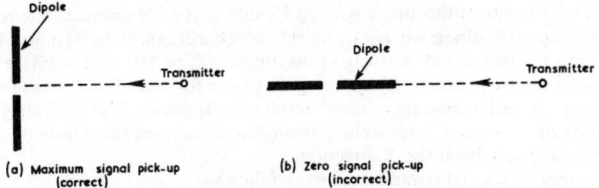

(a) Maximum signal pick-up
(correct)

(b) Zero signal pick-up
(incorrect)

FIG. 6.22 CORRECT AND INCORRECT ORIENTATION OF HORIZONTALLY MOUNTED PLAIN DIPOLE (PLAN VIEW)

only and is rarely met with in the other television bands. Fig. 6.22 shows the correct and incorrect orientation of a horizontally mounted plain dipole.

AERIAL ARRAYS

The directional characteristics of the plain dipole may be considerably improved by positioning additional elements close to it to form an array. The simplest array consists of a dipole and a reflector arranged in H form. A reflector is a conductor slightly longer than the dipole and is placed about a $\lambda/4$ from it as shown in Fig. 6.23. The reflector is not broken at the centre like the dipole but is a continuous conductor. By making the

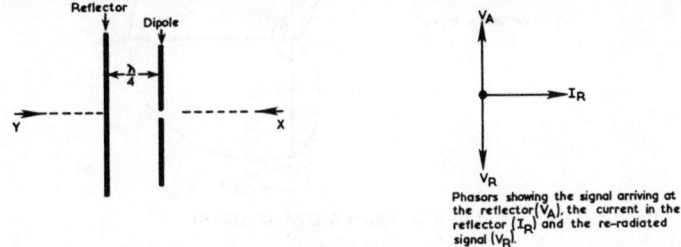

Phasors showing the signal arriving at the reflector(V_A), the current in the reflector (I_R) and the re-radiated signal (V_R).

FIG. 6.23 DIPOLE PLUS REFLECTOR

reflector longer than the dipole it will have an inductive reactance at the frequency to which the dipole is tuned.

The addition of the reflector modifies the polar diagram to that shown in Fig. 6.24, and the reason will be briefly considered. Suppose that a transmission is being received

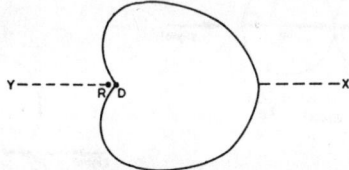

FIG. 6.24 POLAR DIAGRAM OF VERTICALLY MOUNTED DIPOLE PLUS REFLECTOR

from the direction X. The arriving wave will be intercepted by both the dipole AND reflector. However, the signal arriving at the reflector (V_A) will be lagging the signal arriving at the dipole due to the extra $\lambda/4$ the signal must travel to reach the reflector. The signal induced in the reflector will set up a current (I_R) lagging behind V_A by approximately 90° since the reflector is inductive. The reflector reradiates a signal V_R lagging by 90° on the reflector current, i.e. lagging 180° behind V_A. This reradiated signal is picked up by the dipole but lagging V_R by a further 90° because of the $\lambda/4$ spacing. Thus the wave arriving at the dipole from the reflector will be out of phase by 360° or in phase with the direct signal arriving at the dipole. In consequence the two signals augment one another.

When a wave is arriving from the Y direction, the signal arriving at the reflector (V_A) gives rise to a reradiated signal (V_R) as before and 180° out of phase with V_A. The direct signal arriving at the dipole will lag V_A due to the $\lambda/4$ spacing. The reradiated signal arriving at the dipole will lag V_R by 90 which will cancel the 90 lag of the direct signal. Thus the two signals arriving at the dipole will be 180 out of phase with one another and therefore cancel. By applying the above for other directions of approach to the array, the polar diagram of the H aerial may be drawn. The polar diagram now favours the desired signal approaching from X and removes the effects of unwanted signals, particularly from the Y direction.

The important aerial parameters are as follows.

Beam Width

This is a convenient way of measuring the directivity of an aerial. It is the angle subtended by the points at which the received signal power has fallen to half its maximum power or the induced aerial voltage has fallen to 0·707 of its maximum value. For the plain dipole, the beam width $\theta = 90°$ and for the H aerial $\theta = 180°$, see Fig. 6.25.

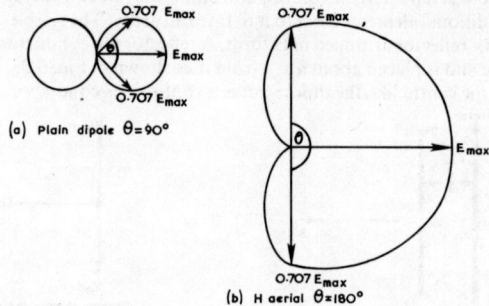

(a) Plain dipole $\theta = 90°$

(b) H aerial $\theta = 180°$

FIG. 6.25   BEAM WIDTH OF AERIAL

Front-to-Back Ratio

This is a ratio of the relative pick up of signal voltage in the forward and backward directions. For the plain dipole it is $E_F/E_B = 1$ or in dB $20 \log 1 = 0\,dB$, see Fig. 6.26. For the H aerial the front-to-back ratio is infinite.

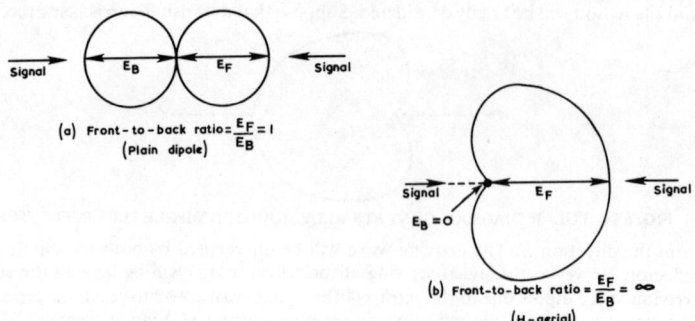

(a) Front-to-back ratio $= \dfrac{E_F}{E_B} = 1$
(Plain dipole)

(b) Front-to-back ratio $= \dfrac{E_F}{E_B} = \infty$
(H-aerial)

FIG. 6.26   FRONT-TO-BACK RATIO OF AERIAL

Aerial Gain

The gain of a receiving aerial is the ratio of power delivered to a matched load to the power delivered by a reference aerial, the field strengths at the location of the aerials being the same.

The reference aerial is either a $\lambda/2$ dipole or an 'isotropic radiator', which will radiate equally in all directions. Such an aerial is not a practical proposition but is a useful concept. It can be shown that the gain of a $\lambda/2$ dipole is about 2 dB better than an isotropic aerial.

The aerial gain of the H aerial is 3 dB over the half-wave dipole.

The high front-to-back ratio of the H aerial combined with its wide acceptance angle enables it to be orientated to reduce the pick up of an unwanted signal arriving from a direction different from the main signal, see Fig. 6.27. By rotating the aerial so that maximum response is away from the direction of the wanted signal, the pick up of the reflected signal E_R is reduced at the expense of some reduction in response to the wanted signal E_W.

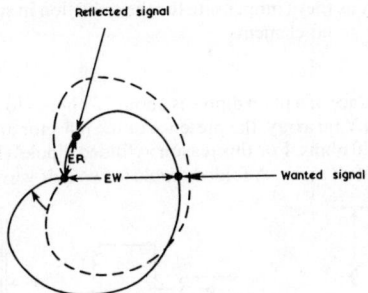

FIG. 6.27   USE OF FRONT-TO-BACK RATIO TO REDUCE PICK-UP OF AN UNWANTED REFLECTED SIGNAL

Adding Directors

To improve directivity additional elements may be added to the H aerial to form a 'Yagi array', named after a Japanese engineer who developed it. The Yagi array forms the basis of nearly all v.h.f. and u.h.f. aerials intended for domestic use, see Fig. 6.28.

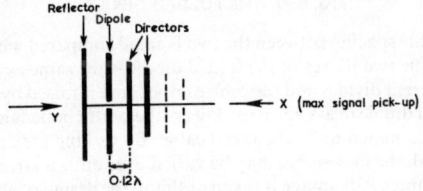

FIG. 6.28   THE YAGI ARRAY

The additional elements, called 'directors', are slightly shorter than the dipole and are placed on the opposite side to the reflector with a spacing of about $0.12\ \lambda$. One or more directors may be used in the array and, like the reflector, they are continuous conductors. The length and spacing of the directors are adjusted so that the reradiated signals from them assist the signal arriving at the dipole when the signal approaches the array from the direction X.

The effect on the polar diagram of adding directors is to increase the forward gain and reduce the beam width as shown in Fig. 6.29. Any number of director elements may be added to provide further increase in gain. The extra yield in gain is about 1.5 dB per element up to five directors but falls off when more are added. As directors are added the polar diagram becomes more difficult to control and side-lobes sometimes

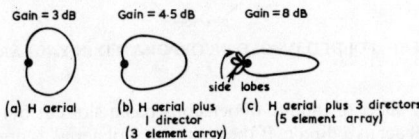

FIG. 6.29   EFFECT ON POLAR DIAGRAM OF ADDING DIRECTORS

appear. In general the design of an aerial array is a compromise between the following features: (a) gain; (b) beamwidth; (c) front-to-back ratio; (d) side-lobe production; (e) bandwidth; and (f) impedance.

The reduction in beam width of a multielement Yagi array is of great assistance in reducing the pick up of reflected signals, particularly in Bands 3, 4 and 5. The higher the frequency of operation the shorter the elements in the array, thus multielement arrays are easier to construct for the higher television bands. A Band 3 array may contain up to 10 directors providing a forward gain of about 11 dB, whereas a high-gain Band 4/5 array may utilise 16 directors giving a gain of about 14·5 dB. The extra directors are useful in a u.h.f. array as they compensate for the reduction in signal pickup due to the shorter length of the aerial elements.

Folded Dipole

The centre impedance of a plain dipole is about 75 ohms. However, when the dipole is incorporated into a Yagi array, the presence of the reflector and directors lowers the impedance to about 20 ohms. For this reason a 'folded dipole' (Fig. 6.30) is often used as the main element of an array. A folded dipole is two half-wave dipoles connected in

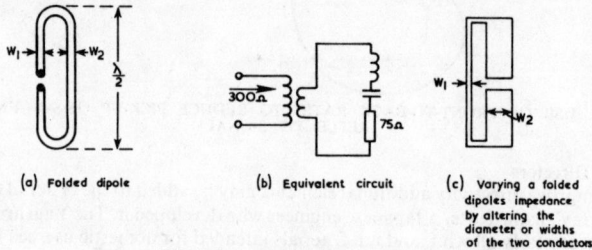

(a) Folded dipole   (b) Equivalent circuit   (c) Varying a folded dipoles impedance by altering the diameter or widths of the two conductors

FIG. 6.30   THE FOLDED DIPOLE

parallel, provided the spacing between the two is small compared with the wavelength. If the diameter of the two halves of the folded dipole is the same as in diagram (a) the available signal current divides and the centre impedance is raised by a factor of four to 300 ohms. A folded dipole in a Yagi array lowers the centre impedance and by suitable design can provide a match to 75 Ω coaxial cable. By making the diameter of the two conductors unequal, the impedance may be varied over quite a large range (greater or smaller than 300 ohms). Advantage is taken of this in the design of arrays having many directors. In some u.h.f. aerial designs the folded dipole is made from a single sheet of metal, suitably cut and bent to provide unequal width conductors as in diagram (c). Fig. 6.31 shows a folded dipole incorporated into a Yagi aerial forming a 6-element array.

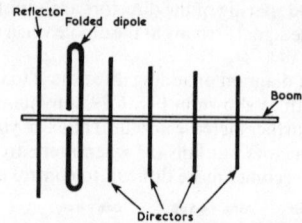

FIG. 6.31   FOLDED DIPOLE INCORPORATED IN YAGI ARRAY

Slot Aerials

If a conducting sheet of infinite dimensions has a slot cut out of it at as in Fig. 6.32(a), the slot will act as a dipole. If the resulting 'slot aerial' is used for transmission and energised at the edges $X-Y$, radiation takes place from either side of the sheet. The

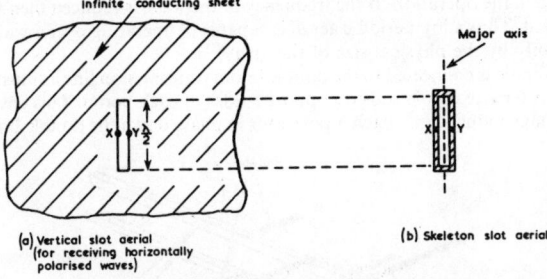

(a) Vertical slot aerial
(for receiving horizontally
polarised waves)

(b) Skeleton slot aerial

FIG. 6.32 SLOT AND SKELETON SLOT AERIALS

polar diagram of radiation is similar to that of a plain dipole, except that the plane of polarisation of the wave will be turned through 90°. In practice the sheet can be of finite dimensions and can be used for receiving as well as transmitting. A vertical slot is used for receiving horizontally polarised waves and a horizontal slot for vertically polarised waves. Slot aerials have the advantage that if installed in a loft the longest dimension is the width of the conducting sheet when dealing with vertically polarised waves and not the height, which is most useful for a Band 1 loft aerial.

The 'skeleton slot' aerial is a development of the slot aerial. By gradually reducing the surrounding conducting sheet, the physical size is reduced whilst giving similar results, see diagram (b). The sheet now consists of a thin metal rim surrounding the slot which in practice may be made from aerial tubing. Connection to the slot is made at X–Y where the impedance is higher than an ordinary dipole (about 500 ohms). Because of the higher impedance a 'matching unit' is required for connecting to 75 Ω cable. The skeleton slot may be incorporated into a directional array, but the directors and reflector will be mounted at 90° to the major axis of the slot. The skeleton slot principle is used in some Band 4/5 arrays and Band 3 aerials.

Log-Periodic Aerial

This is a development of the Yagi array and consists of a series of dipoles having lengths and spacings which increase in a logarithmic manner as shown in Fig. 6.33. All the dipoles are connected to a common twin transmission line, which in practice is

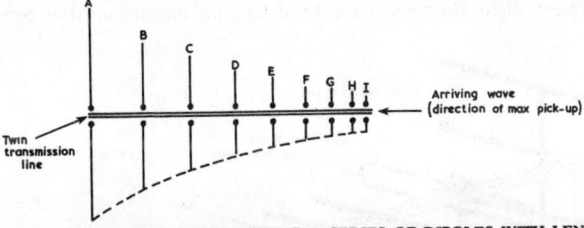

FIG. 6.33 'LOG PERIODIC' PRINCIPLE USING A SERIES OF DIPOLES WITH LENGTH AND SPACING VARYING IN A LOGARITHMIC MANNER

formed by a double boom. Each dipole has shorter dipoles on one side and longer dipoles on the other and so behaves like a Yagi aerial.

Consider a wave arriving as shown having a frequency which corresponds to the resonant frequency of dipole C. As the wave moves along the array it first meets the shorter non-resonant dipoles I, H, G, F, etc. to which it gives up energy. As the wave progresses it gives up more and more energy to the intercepting dipoles until it reaches dipole C where maximum energy is yielded. The next element that the wave 'sees' is dipole B which behaves like a conventional Yagi reflector, thus the rest of the array to the left of dipole B contributes little to the operation of the aerial. As the frequency of the arriving wave is reduced, making, say, dipole B the resonant one, more of the array is involved in the operation. If the frequency of the wave is reduced then fewer dipoles are involved. Thus a log-periodic aerial is capable of operating over a wide bandwidth (limited only by the physical size of the array).

Each dipole is connected to the double boom transmission line as illustrated in Fig. 6.34 with alternate dipole rods on opposite sides of each boom. This ensures that the signal voltage induced into each dipole adds in phase down the double boom to which

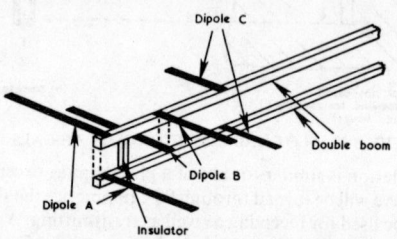

FIG. 6.34 CONNECTION OF THE DIPOLE RODS TO THE DOUBLE BOOM IN A LOG PERIODIC AERIAL

the feeder cable is connected. The log-periodic aerial may be mounted vertically or horizontally according to the wave polarisation and the number of elements varied according to the field strength available at the aerial location.

Like the Yagi array, the log-periodic aerial has a high front-to-back ratio and good directivity and so is suitable for rejecting unwanted reflected signals or interference. Because of its wide bandwidth capabilities this type of aerial may be designed to cover the whole of Band 3 in a v.h.f. array or any group in the u.h.f. band. At least one aerial manufacturer offers a u.h.f. log-periodic aerial which covers all of the u.h.f. channels. This feature is particularly useful when the aerial is to be used in different locations, *e.g.* with a receiver fitted in a caravan.

U.H.F. Aerial Arrays

Aerials for use at u.h.f. are based on the Yagi array principle but often differ in detail. A typical 10-element array is shown in Fig. 6.35. This uses a folded dipole as the main element, eight directors and a 4-rod reflector unit (counted as one element)

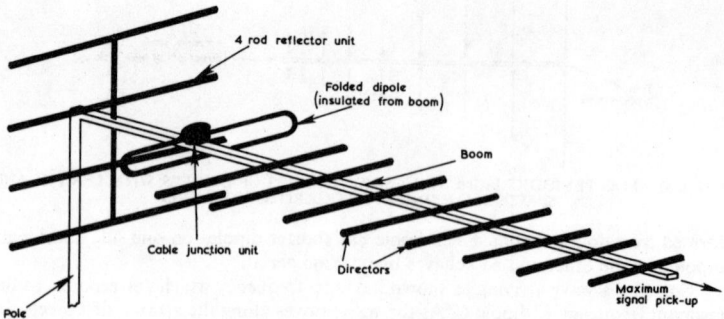

FIG. 6.35 10-ELEMENT U.H.F. AERIAL

mounted at right angles to the boom. A multi-rod reflector unit is a common feature with a u.h.f. array and is used to maintain a high front-to-back ratio over the full bandwidth of the aerial. Some aerial manufacturers mount the reflector unit at an angle to the boom as a corner unit which gives uniform channel performance, ensuring a high front-to-back ratio and at the same time cutting down wind resistance. In other designs, the reflector is made from a metal sheet with slotted apertures.

Typical parameters for three group A aerials are:

NO. OF ELEMENTS	USE	FORWARD GAIN	FRONT/ BACK RATIO	BEAM WIDTH
6	Areas of good signal strength	9·5 dB	25·2 dB	50
13	Medium range reception	13·0 dB	27·2 dB	38
18	Long range reception	14·7 dB	30·7 dB	32

U.H.F. aerials must have a bandwidth of at least 88 MHz for the reception of the four local programmes, and in selecting an aerial for a particular group its performance over the channels concerned is important. The solid curve of Fig. 6.36 shows what may occur with some aerial designs where the gain falls off at the extremes of its bandwidth

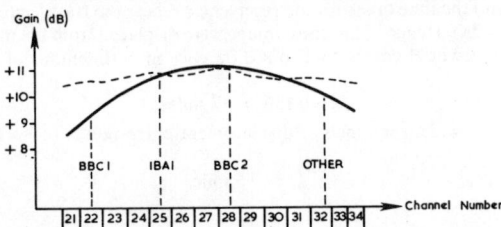

FIG. 6.36   VARIATION IN AERIAL GAIN OVER THE 11-CHANNEL BANDWIDTH REQUIRED FOR RECEPTION OF THE FOUR LOCAL PROGRAMMES (GROUP A AERIAL)

coverage. This may cause unwanted luminance and colour variations between different programmes. The dotted curve shows the response of an aerial of good design where the variations are within 1 dB over the full group of channels.

An outdoor aerial mounted clear of any obstructions and above roof-top always provides the stronger signal. However, some may be obliged to use an indoor aerial. Because of its small dimensions, a u.h.f. aerial array may be conveniently fitted in the loft space but the signal is generally about 10 dB lower than the 'above roof' mounted array (about 5 dB is due to loss of height and 5 dB attenuation due to the roof material). Apart from this there may be a further deterioration due to the presence of electrical wiring, pipes, water tanks, etc. close to the aerial which may upset the polar response and cause reflections resulting in ghost images. Finding the optimum position is a matter of trial and error and considerable patience is required to find a position giving the best signal-to-noise ratio and freedom from ghost images. Always stand behind the reflector and well clear of the array when assessing the signal value in a particular position.

Room or 'set-top' aerials are available in a variety of designs. Some consist of a small Yagi array with provision for adjusting the array to suit vertically or horizontally polarised signals. Others use a pair of circular or square elements arranged as a main element-reflector system. Although there is a measure of directivity with these simple arrays the performance is greatly influenced by positioning. A 'set-top' aerial may suffer a signal loss of about 25 dB compared with the 'above roof' aerial. Most of this is due to loss of aerial height, but some is due to attenuation through outside walls, party walls or partition walls. Reflections from objects and people inside the room tend to introduce variations in luminance and colour which can be most disturbing when people move about.

MULTIPATH RECEPTION

Signals reflected by land masses, buildings, gas-holders, etc. may cause ghost images to appear on the screen of the receiver. Fig. 6.37 shows two possible reflected signal paths. In both cases a 'direct' and a 'reflected' signal arrive at the receiving aerial.

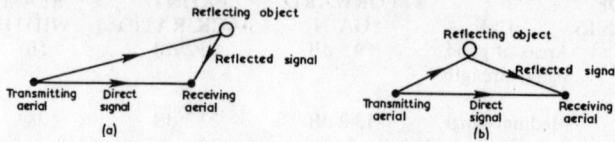

FIG. 6.37   REFLECTIONS CAUSING GHOST IMAGES

Since the direct and reflected waves travel over different path lengths, there will be a time difference between the receipt of the two signals at the receiver. In addition there will be an amplitude difference, with the direct signal generally being the larger of the two. This may not always be so as the direct signal may suffer a greater attenuation than the reflected signal depending upon the intervening terrain in the two signal paths. Because the reflected signal arrives at a later instant it will give rise to a ghost image.

The speed of travel of radio waves is 186,000 miles per second or 0·186 miles per micro-second and the time taken for the receiver c.r.t. beam to travel across the screen is 52 μs (left to right). Hence, if the ghost image were displaced from the main image by a complete line, it would correspond to a difference in path lengths of

$$52 \times 0\cdot186 \simeq 9\cdot7 \text{ miles.}$$

With a screen 18″ wide, each inch of displacement corresponds to

$$\frac{9\cdot7}{18} \simeq 0\cdot5 \text{ mile.}$$

This might be worth remembering as it may help in determining what is causing the ghost image.

Effect of Ghost Images on Monochrome Reception

Ghost images may be negative or positive depending upon whether the signals are in phase or in antiphase. Fig. 6.38(a) shows the direct and reflected signals in phase at the aerial during a 625-line monochrome transmission. The signal fed to the receiver is the resultant of the two signals and as shown produces a negative ghost, *i.e.* the resultant signal envelope increases towards black level. On a monochrome receiver the white bar is accompanied by a black bar ghost which is displaced to the right. Diagram (b) shows the result when the two signals are in antiphase and the resultant signal causes a white bar ghost image (but less intense) displaced to the right of the main image. Since the wavelength of the received signals on u.h.f. is extremely short, quite a small movement of the aerial (due to wind or slight rotation during erection) will cause the ghost image to change from positive to negative or *vice versa*.

The removal or reduction of the effects of ghost images arising from the reflecting objects in Fig. 6.37 can often be done by suitable choice and positioning of the aerial array. Fig. 6.39(a) shows how a directional array may be used to deal with a reflected signal originating from an object at the back of the aerial. Here the reflected signal pick up is diminished by the high front-to-back ratio of the aerial. By using an even more directional array as in diagram (b), a forward type reflected signal may be diminished by the narrow acceptance angle of the aerial array.

Metal objects, *e.g.* other aerials, poles, metal gutters, etc. close to an aerial may upset the polar response which can lead to ghost signals, so a site should be chosen which is well clear of such objects. Short-term ghosting by nearby objects often gives rise to 'soft' vertical edges in the picture.

Another cause of reflections is from flying aircraft ('aircraft flutter'). Waves reflected from a moving aeroplane give rise to a varying time difference between the

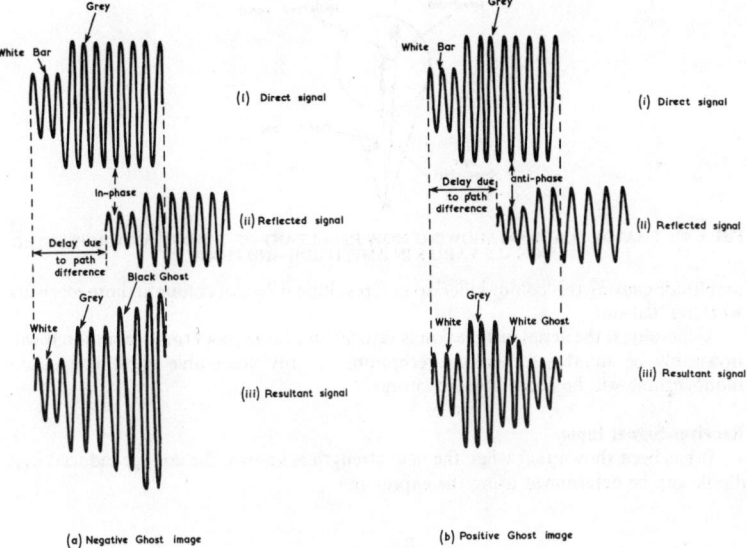

(a) Negative Ghost image   (b) Positive Ghost image

FIG. 6.38 PRODUCTION OF POSITIVE AND NEGATIVE GHOST IMAGES DUE TO REFLECTED SIGNAL

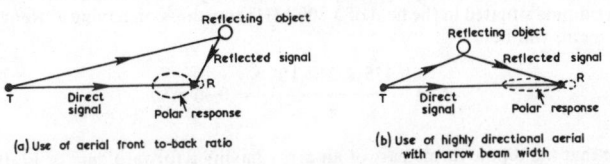

(a) Use of aerial front to-back ratio   (b) Use of highly directional aerial with narrow beam width

FIG. 6.39 USE OF POLAR RESPONSE TO MINIMISE PICK-UP OF REFLECTED SIGNAL

receipt of the direct and reflected signals, which causes large changes in the amplitude of the resultant signal and produces a 'pumping' effect on the picture. Aircraft flutter takes place at a few cycles per second but the effects depend upon the speed of the aircraft and whether it is moving down the beam or across it. From the aerial point of view there is little that can be done to avoid this effect.

Effect of Ghost Images on Colour Reception

Reflected signals adding to the direct signal produce the same effects in the luminance channel of a colour receiver as they do in a monochrome receiver. As this is the most dominating effect in a colour receiver, the results are similar to those obtained in monochrome reception with the ghost image (positive or negative) displayed to the right of the main image on the screen.

The resultant signal of the direct and reflected waves not only varies in amplitude but also in phase, see Fig. 6.40. We have seen that 'phase' is very important when dealing with chrominance signals. However, because of the nature of the PAL signal a change in phase of the received chrominance signal does not cause a change in hue but does result in a variation of the colour saturation. Thus, during periods of ghost signal reception there will be some change in the saturation of the colours in the picture. It is possible for overlap to occur between the direct and reflected burst signal components during periods of strong short-term reflections. This may reduce the burst signal

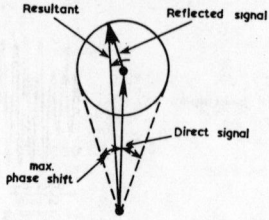

FIG. 6.40 PHASOR DIAGRAM SHOWING HOW RESULTANT OF 'DIRECT' AND 'REFLECTED' SIGNALS VARIES IN AMPLITUDE AND PHASE

amplitude causing the 'colour killer' to fail, resulting in loss of colour in those receivers working 'flat-out'.

Generally, if the aerial installation is satisfactory for monochrome reception it will invariably be suitable for colour reception, *i.e.* any acceptable ghost images on monochrome will be acceptable on colour.

Receiver Signal Input

It has been shown that when the field strength is known, the voltage induced in a dipole can be determined using the expression

$$E = \frac{\lambda}{\pi} e.$$

When the dipole forms part of an array and the gain of the array is known, the voltage is then increased appropriately by the gain of the array. For example, the e.m.f. of a half-wave dipole situated in the field of a 800 MHz transmission having a strength of 3 mV per metre will be

$$E = \frac{0 \cdot 375 \times 3 \times 10^{-3}}{\pi} \simeq 0 \cdot 36 \text{ mV}.$$

Suppose that the dipole forms part of an array having a forward gain of 12 dB. The aerial array e.m.f. (E_a) will be

$$4 \times 0 \cdot 36 \text{ mV} = 1 \cdot 44 \text{ mV}$$

(Note that 12 dB represents a voltage ratio of 4:1).

Not all this voltage, however, reaches the receiver. In the ideal matched condition (see Fig. 6.41) where the receiver load and aerial impedances are the same, the voltage across the input terminals of the receiver will be $E_a/2$, i.e. $1 \cdot 44/2$ mV = $0 \cdot 72$ mV. A

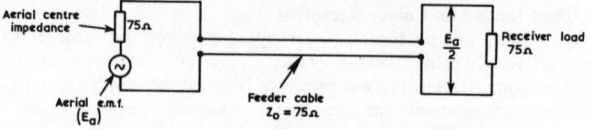

FIG. 6.41 MATCHED AERIAL-RECEIVER SYSTEM

signal level of this order at the receiver terminals will probably be adequate for the majority of colour receivers. Generally, if the signal is not less than about 500 μV the noise grain on the screen will not be too objectionable during colour reception. A lower signal level can usually be tolerated on monochrome as the size of the noise grain is smaller.

In fringe or shadow reception areas where the signal is weak resulting in noisy pictures and sound, a preamplifier or 'booster' amplifier may be used. The best arrangement is to place the amplifier at the top of the mast where the signal is the largest, so that noise introduced by the amplifier will have the least effect. U.H.F. masthead amplifiers are available for operation over a particular group of channels, *e.g.* group A, B or C/D and this arrangement minimises the chances of amplifying unwanted signals. These amplifiers provide a gain of about 11—15 dB depending upon the aerial group and must be fully weather-proofed for mounting on the mast. The d.c. supply to the masthead amplifier (transistor type) may be obtained from a separate mains power unit mounted near to the receiver and fed to the amplifier *via* the aerial down-lead. An L.C. filter will be incorporated to separate the d.c. and signal components.

Attenuators

If the signal strength is high the receiver may be overloaded and an attenuator is required between the aerial down-lead and the receiver input socket. Either of the two attenuator circuits shown in Fig. 6.42 may be used. In designing these circuits for a given attenuation, the impedance seen 'looking' into the attenuator from either

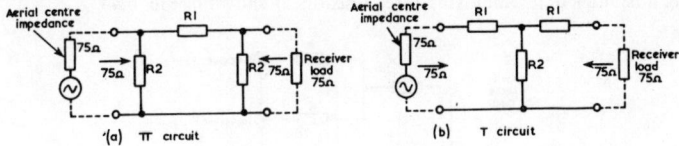

FIG. 6.42   ATTENUATOR CIRCUITS

direction must be the same (usually 75 Ω for television purposes). With the π-type circuit the shunt arms (R_2) have the same value, whereas for the T-type the series arms (R_1) have the same resistance value. The value of the resistances are:

π-type:

$$R_1 = \frac{R(A^2 - 1)}{2A} \quad ; \quad R_2 = \frac{R(A + 1)}{(A - 1)}$$

T-type:

$$R_1 = \frac{R(A - 1)}{A + 1} \quad ; \quad R_2 = \frac{R(2A)}{A^2 - 1}$$

where R is the nominal impedance of the system, *e.g.* 75 Ω and A is the ratio of input-to-output signals.

For example, suppose $R = 75$ Ω and an attenuation of 18 dB (8:1) is required.

For the π-type:

$$R_1 = \frac{75(8^2 - 1)}{2 \times 8} = 295 \cdot 3 \text{ Ω} \quad (300 \text{ Ω})$$

$$R_2 = \frac{75(8 + 1)}{8 - 1} + 96 \cdot 4 \text{ Ω} \quad (100 \text{ Ω})$$

For the T-type:

$$R_1 = \frac{75(8 - 1)}{8 + 1} = 58 \cdot 3 \text{ Ω} \quad (56 \text{ Ω})$$

$$R_2 = \frac{75(2 \times 8)}{8^2 - 1} = 19 \cdot 0 \text{ Ω} \quad (18 \text{ Ω})$$

Since the values are not critical, the nearest standard preferred value resistors may be used (shown in brackets). There is basically no difference in the two attenuator circuits, but one will be more suitable because of the resistances available depending upon the attenuation required. The circuits shown are for an unbalanced system, *e.g.* with coaxial feeder. In a balanced feeder, the same arrangement may be used except that the calculated R_1 is split, half being placed in each lead thereby forming symmetrical π and T circuits.

Although these simple attenuators may easily be constructed, it should be noted that the self-inductance of the resistors and stray capacitances may, at u.h.f., result in an attenuation quite different from the calculated values. Commercial attenuators (in-line attenuators) are usually designed for use at v.h.f. and u.h.f. They are usually of coaxial type construction and are available with fixed attenuation levels, *e.g.* 6 dB, 12 dB, 18 dB, etc.

Splitter Units

It may sometimes be necessary to operate more than one receiver from a common aerial system, *e.g.* in households which have a colour receiver in one room and a monochrome receiver in another. The receiver inputs cannot be connected in parallel without upsetting the matching. Two receivers may be operated from a single aerial using a splitter unit' employing three resistors as shown in Fig. 6.43.

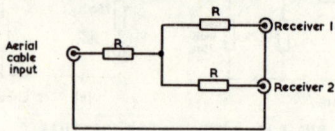

FIG. 6.43   TWO-WAY SPLITTER UNIT

The value of each resistor is the same and may be calculated from

$$\frac{n-1}{n+1} Z_o,$$

where n is the number of receivers and Z_o is the nominal impedance of the system (normally 75 ohms). Thus for a two-way splitter

$$R = \frac{2-1}{2+1} \times 75 = 25 \text{ ohms.}$$

To operate three receivers (using four resistors) each resistor would be

$$\frac{3-1}{3+1} \times 75 = 37 \cdot 5 \text{ ohms.}$$

In both cases the nearest preferred value resistors would be used. If an output is not required it should be terminated with a 75 Ω resistor so as to maintain correct matching.

Operating receivers in this way obviously reduces the signal to each receiver. With two receivers, there is a 6 dB loss of signal voltage at each output compared with the available input to the splitter unit. For three receivers the loss is about 10 dB. Thus when a large number of receivers are to be supplied from a common aerial system more elaborate arrangements are necessary and distribution amplifiers become essential.

Feeder Extensions

To operate a single receiver in one of two locations, the feeder may be extended as shown in Fig. 6.44. In position A the extension feeder is not in use. In position B, the cable from A to B is plugged into the socket at A. More than two operating positions

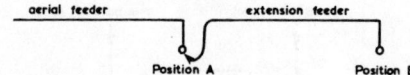

FIG. 6.44   EXTENSION OF FEEDER CABLE TO PROVIDE TWO OPERATING POSITIONS FOR
A RECEIVER

may be provided simply by repeating the extension arrangement between position A
and B.

Combining and Dividing Units

If two aerials are to be combined in a common down-lead, a 'combiner' or
'diplexer' must be used. The aerials cannot be connected directly in parallel as the
matching would be upset; also there may be a problem with ghost signals as each aerial
will pick up some signal to which the other aerial is resonant. It may be necessary to
combine a Band 2 v.h.f. radio aerial with a Band 4/5 u.h.f. television aerial into a
common down-lead; the arrangement shown in Fig. 6.45 may be used. The combiner
consists of two T-type filters, one being a high pass and the other a low pass. Each aerial

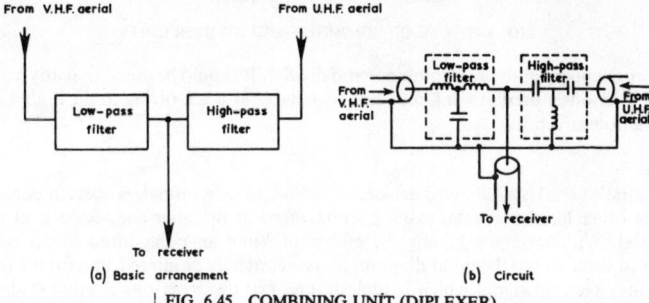

(a) Basic arrangement   (b) Circuit

FIG. 6.45   COMBINING UNIT (DIPLEXER)

thus 'sees' the feeder cable *via* its associated filter but the aerials do not 'see' one
another. Fig. 6.46 shows the response of a typical v.h.f./u.h.f. combiner with a
crossover frequency of about 350 MHz.

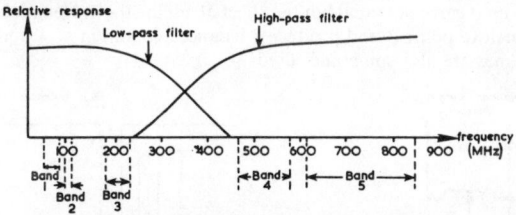

FIG. 6.46   RESPONSE OF V.H.F./U.H.F. DIPLEXER

At the receiving end of the cable, the Band 2 and Band 4/5 signals must be
serparated from one another and fed to the respective receivers. Since a combiner is
reversible in action it may be used to separate the signals in the down-lead as shown in
Fig. 6.47. It is now called a 'divider': the essential difference between a combiner and a
divider is one of construction. A combiner has to be rugged and weather-proof since it
is normally mounted outdoors, whereas a divider is normally mounted indoors and has
a more pleasing appearance.

All combiners and dividers introduce some loss, but this is usually small in a well-
designed unit. Whether the arrangement shown in Fig. 6.47. is used or separate down-
leads from the two aerials depends upon the installation. If the feeder run is short,
separate down-leads may be the better arrangement; but with a long feeder run it may

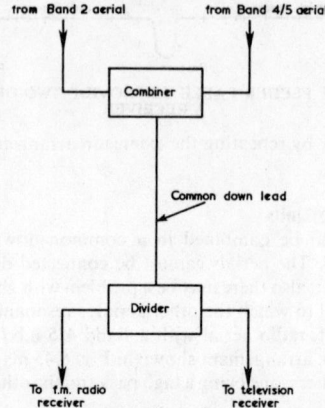

FIG. 6.47 USE OF COMBINER AND DIVIDER UNITS

be more economical to use a combiner and divider. It should be noted that any existing feeder previously used at v.h.f. may not be suitable at u.h.f. owing to the higher cable attenuation at u.h.f.

Balun

A dipole aerial is a balanced device, *i.e.* neither of its terminals is at earth potential. On the other hand a coaxial cable is unbalanced as its outer conductor is at earth potential. With receiving aerials the effects of using an unbalanced feeder with a balanced aerial are (i) the polar diagram of the aerial may be altered; and (ii) the feeder cable may pick up signal which is undesirable. For these reasons a balun (**bal**anced **un**balanced) may be used.

There are a number of ways of obtaining a balun action. One would be to use an auto-transformer as shown in Fig. 6.48(a) with a 1:2 ratio between the coaxial feeder and the aerial. Another method shown in diagram (b) uses a $\lambda/4$ length of dummy coaxial cable shorted at each end and connected as shown. Without this length of cable, point Y would be at earth potential but the effect of adding the $\lambda/4$ length of cable is to transfer the earth to point Z and produce a balanced condition at X and Y. Baluns based on $\lambda/2$ lines are also sometimes used.

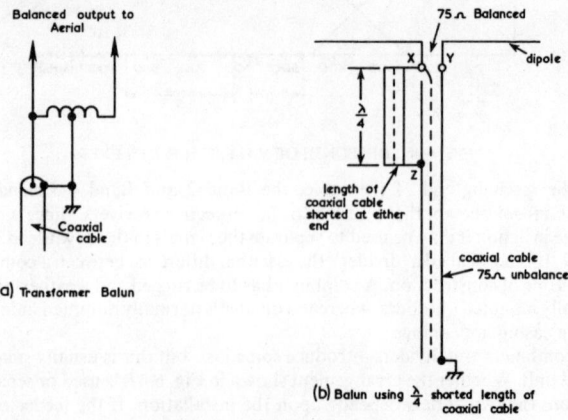

FIG. 6.48 BALUN FORMS

CHAPTER 7

RECEIVER BLOCK DIAGRAMS

A good block diagram enables the technician to quickly become familiar with the path of the signal and its processing in various parts of the complete circuit. This is most important in a complex receiver where a number of integrated circuits are used to perform many circuit functions, some of which are not directly related to the position of the i.c. in the circuit diagram layout. Unfortunately, a block schematic is not always included with the service information and one has to fall back on basic block diagrams memorised earlier.

Basic block diagrams showing the main circuit functions for monochrome and colour receivers are given in Figs. 7.1 and 7.2. These diagrams (or similar ones) should be studied before and during learning about the operation of the various circuits.

The general format of the monochrome receiver has changed very little since the introduction of 625-line operation in this country, although many improvements have been made with regard to the sharpness, brightness and stability of the black-and-white picture and the quality of the sound. A simplified description of the basic monochrome block diagram was given in Chapter 2. At one time 'direct' sync. was used for the line timebase in the majority of receivers, with 'flywheel' sync. reserved for models intended for fringe area reception. Flywheel sync. is now a common feature and improvements in design results in more consistent triggering of the line timebase. At the time of writing some current receivers are hybrid (a mixture of valves and transistors) but the tendency is towards solid-state electronics throughout, with integrated circuits taking over many of the circuit functions previously performed by discrete components. The portable version of the monochrome receiver is now regarded by many as the 'second set' to be used in the kitchen or elsewhere in a house or caravan and here, perhaps, lies the future of the monochrome receiver.

The first point to be made concerning the basic colour receiver block diagram is its

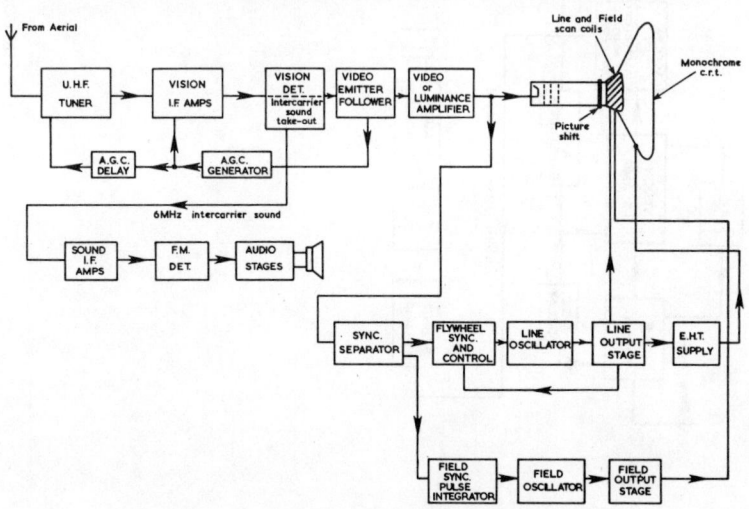

FIG. 7.1 BASIC MONOCHROME RECEIVER BLOCK DIAGRAM

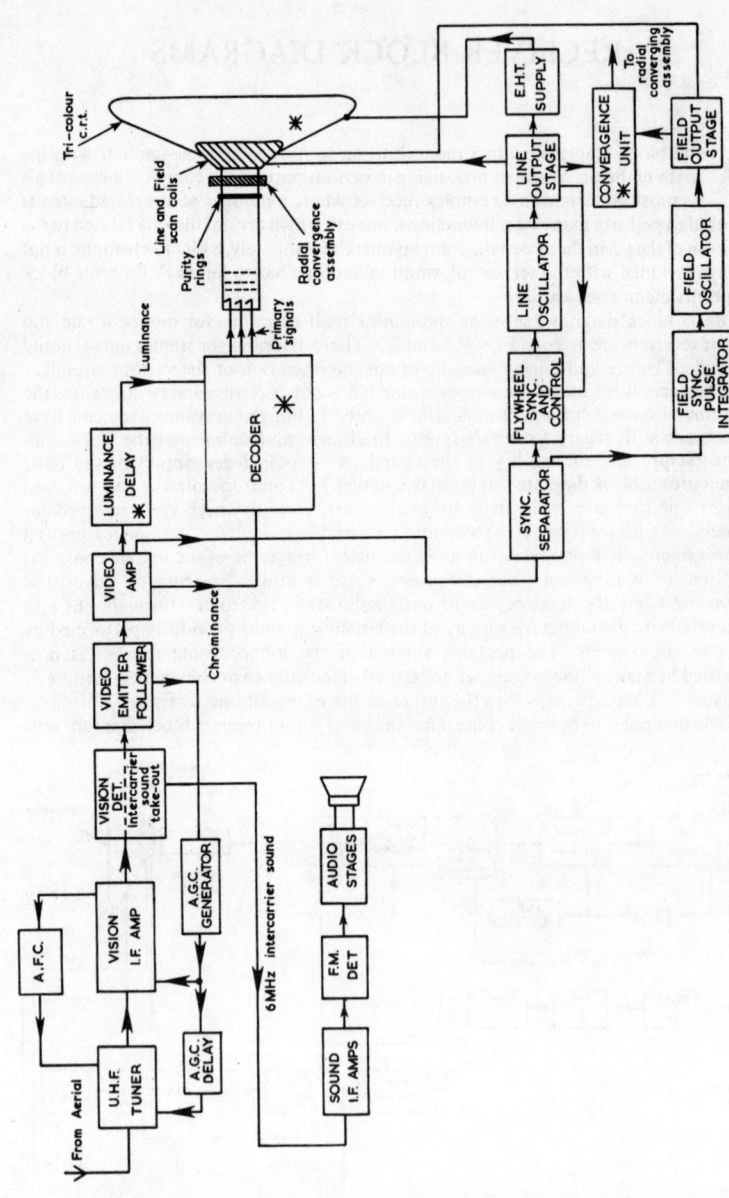

FIG. 7.2 BASIC COLOUR RECEIVER BLOCK DIAGRAM

similarity to the monochrome receiver schematic. Many of the stages are identical, *e.g.* tuner, vision i.f. amps, vision det. sound section, sync. separator and timebases, etc. The main differences between the two receivers have been marked with an asterisk in Fig. 7.2.

A tricolour c.r.t. is required to produce the colour image which is formed by energising phosphors emitting primary coloured lights with the aid of three separate electron beams. These beams have to be kept in register or CONVERGED during the scanning process and is carried out *via* the radial convergence assembly (for a delta gun c.r.t.) fed with suitable currents from the CONVERGENCE UNIT. With a PRECISION IN LINE colour c.r.t. the problems of convergence are reduced and a different arrangement is used for ensuring that the three beams are converged. To deal with the colouring or chrominance signal, a decoder is required which carries out many functions but is shown simply as a single block in Fig. 7.2. The output from the decoder drives the guns of the colour tube with primary signal waveforms to produce the necessary three colour images. A colour receiver also requires a luminance delay line which ensures that the luminance information is registered at the same time as the chrominance information. Without this delay the colour would be displayed to the right of the luminance information on the screen of the c.r.t.

The monochrome receiver may be regarded as being a simplified version of the colour receiver where the natural tone colour picture is replaced by an unnatural but acceptable monochrome image. Thus the monochrome receiver does not require the decoder, convergence and luminance delay of the colour receiver and the tricolour c.r.t. is replaced by a simpler monochrome tube.

Decoder Block Diagram

A basic block diagram of a colour receiver decoder is shown in Fig. 7.3 and a brief description of the function of the various stages will now be given. The waveforms on the diagram illustrate the operation during reception of standard colour bars.

The output of the vision detector contains the composite picture signal, *i.e.* luminance, chrominance, subcarrier burst and sync. pulses which are fed *via* an emitter-follower stage block (1) to a video amplifier block (2). In this stage the chrominance and burst signals are extracted from the composite waveform using a fairly low Q tuned circuit. This tuned circuit has a bandwidth of approximately ± 1 MHz centred on 4·43 MHz and embraces the chrominance and burst signals. The same tuned circuit may also be used to prevent the chrominance and burst from reaching the luminance channel. Thus the output of the video amplifier contains luminance and sync. signals only which are passed *via* the luminance delay stage to block (23) in the decoder (about which more will be said later).

The chrominance signal fed to block (4) in the decoder contains frequencies extending from approximately 3·43 MHz up to 5·43 MHz and so contains a small amount of luminance signal energy extracted, together with the chrominance signal in block (2). However, the luminance signal energy has an insignificant effect upon the decoder operation. Burst and chrominance are amplified in block (4) which has two outputs. One is fed to a further stage of amplification block (11), and the other to block (5) which will be dealt with first.

(1) Burst Signal Path

It is required to separate the burst signal from the chrominance information so that the burst may be used to synchronise the crystal oscillator, block (8), the output of which is used to recreate the subcarrier (suppressed at the transmitter). Without this subcarrier the chrominance information cannot be detected. The burst gate stage is fed with gating pulses and operates only when the pulses are present. These pulses, which are usually derived from delayed line flyback pulses, having a timing and duration that corresponds to the period of the burst signal, *i.e.* the back porch period. Thus, only the burst signal appears at the output of block (5). The burst signal is then fed to the phase

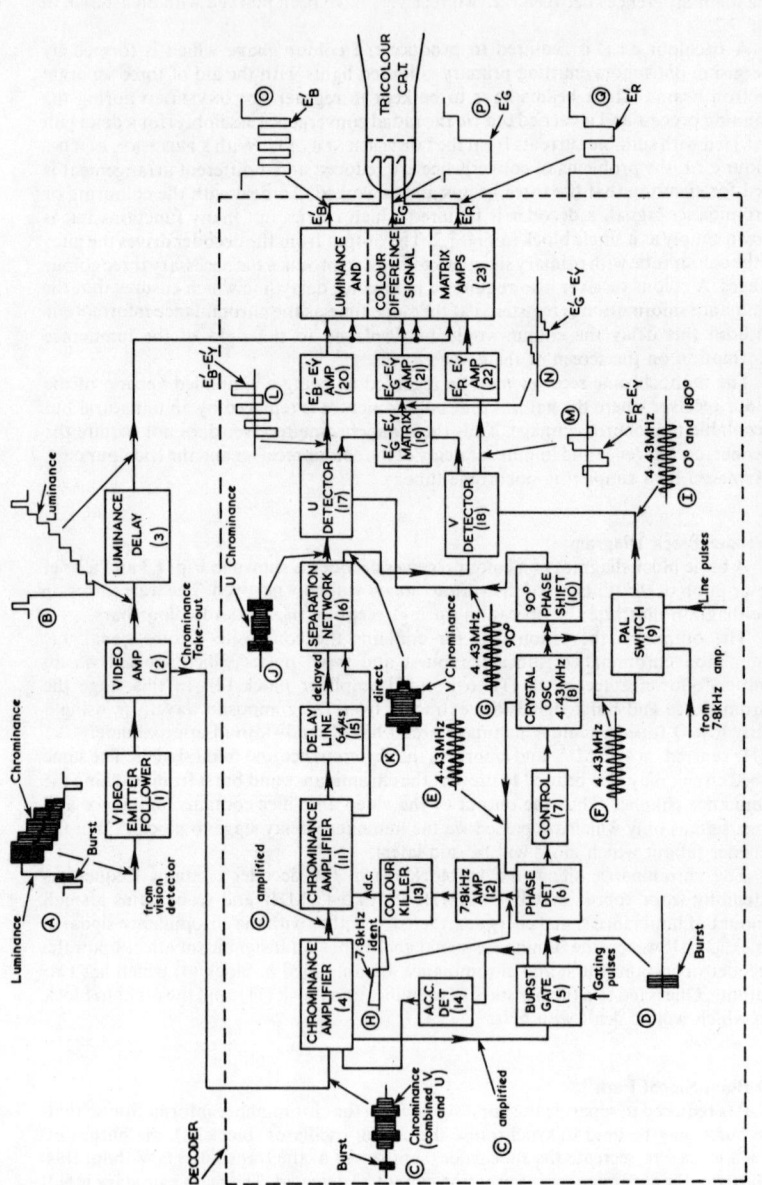

FIG. 7.3　BASIC COLOUR RECEIVER DECODER BLOCK DIAGRAM

detector, block (6), which also has another input (a continuous sine-wave) from the crystal oscillator. The frequency and phase of the burst and oscillator inputs are compared in the phase detector. If there is an error, a correction voltage voltage is fed *via* the control stage (7) which corrects the frequency and phase of the subcarrier oscillations.

(ii) Crystal Oscillator Output

A subcarrier oscillation is now positively locked to the burst signal and the output of the oscillator is supplied to the two chrominance signal detectors [blocks (17) and (18)]. Block (10) provides a 90° phase shift so that detection of the U signal component (which is at 90° to the V signal component) takes place correctly. The subcarrier input to the V detector comes from the PAL switch (9). The purpose of this stage is to reinvert the V signal component on alternate lines to deal with the PAL signal (remember the V signal component is reversed on alternate lines at the transmitter). Block (9) is really two separate stages. It contains a bistable oscillator operating at 7·8 kHz which is triggered by 15·625 kHz line pulses (note that a bistable oscillator divides by two). The output of the bistable oscillator is used to reverse the phase of the subcarrier input block (8) on alternate lines. This is often achieved by using the bistable output to alternately cause one of a pair of diodes to conduct, thereby passing normal (0°) and inverted (180°) subcarrier to the output. To keep the PAL switch in step with the V signal alternations at the transmitter, block (9) is fed with a 7·8 kHz locking input from the 7·8 kHz amplifier block (12). This stage is fed with the 'identification signal' derived at the output of the phase detector due to the effects of the 'swinging burst'. The identification signal is an approximate square wave and the purpose of block (12) is to amplify the fundamental component of 7·8 kHz.

(iii) Chrominance Signal Path

Return now to block (4) and consider the other output of the first chrominance amplifier which is fed to block (11). The gain of the first chrominance amplifier is normally controlled by an a.c.c. (automatic chrominance control) voltage derived by rectifying the burst signal in the a.c.c. detector of block (14). The a.c.c. voltage maintains a constant burst signal amplitude by controlling the gain of block (4) regardless of changes in amplitude of the signal input to the decoder. Block (11) provides further amplification of the chrominance signal to drive the delay line in block (15). The second chrominance amplifier is operational only during colour transmissions, *i.e.* on monochrome transmissions the stage is cut off. This is achieved by providing the forward bias for this transistor amplifying stage from the colour killer stage. Block (13) rectifies the output of the 7·8 kHz amplifier and the resulting d.c. is used to bias on block (11). As there'is no output from the 7·8 kHz amplifier during monochrome transmissions (no swinging burst present), the chrominance channel is cut off or 'killed'. This prevents coloured noise from spoiling the monochrome picture. The burst signal present at the input to block (11) is not required and may be removed by applying a suitable gating pulse to this stage to cut it off during the burst period.

The amplified chrominance signal output of block (11) feeds the delay line (15) and the separation network (16). The delay line provides the necessary line period delay of 64 μs so that an electrical average of adjacent transmitted lines of chrominance information may be taken. To produce this comparatively long delay, an ultrasonic delay line is used. In the separation network the 'delayed' and 'direct' inputs are electrically averaged and at the same time the V and U components of the signal are separated from one another as explained in Chapter 4.

Separated U and V signal components are then fed to their respective detectors (17) and (18) which also receive subcarrier inputs as previously described. These stages synchronously detect the V and U chrominance signals to provide colour-difference signals at their outputs.

(iv) Colour-Difference Signal Path

The $E_B' - E_Y'$ and $E_R' - E_Y'$ colour-difference signals (0—1 MHz) appearing at the outputs of the detectors are fed to blocks (20) and (22) where they are amplified. Suitable portions and polarities of the $E_B' - E_Y'$ and $E_R' - E_Y'$ signals are also fed to the $E_G' - E_Y'$ matrix where they are combined to form the $E_G' - E_Y'$ signal. Block (21) provides amplification of the $E_G' - E_Y'$ signal. De-weighting of the signals is normally carried out at the detector outputs by adjusting the gains of the $E_B' - E_Y'$ and $E_R' - E_Y'$ channels (the gain of the $E_B' - E_Y'$ channel is increased relative to the $E_R' - E_Y'$ Channel).

In block (23) which contains three separate amplifying stages, the colour-difference signals from blocks (20), (21) and (22) are matrixed with the luminance signal to provide the primary signals E_R', E_B' and E_G'. These signals are fed to the cathodes of the tricolour tube to drive the three guns (primary signal drive). On monochrome, when there are no colour-difference signals, the drive to the three guns consists of the luminance signal only.

Variations in decoder design are to be met with in practice and there will be features not shown in this basic schematic. Integrated circuits are now used for many of the decoder functions but these are not suitable for illustrating the basic decoder operation.

CHAPTER 8

THE U.H.F. TUNER

THE operation of the various stages common to both monochrome and colour receivers will now be considered, commencing with the u.h.f. tuner.

The purpose of the receiver tuner is to select and amplify the desired programme signal and to change its frequency to the receiver i.f., following normal superhet receiver principles. The tuner contains r.f. and frequency-changer stages (see Fig. 8.1) and for operation over Bands 4 and 5 a coverage of 470–854 MHz is required.

An r.f. amplifier stage is required to increase the signal level at the mixer input above the critical noise level, *i.e.* to improve the signal-to-noise ratio. This is important

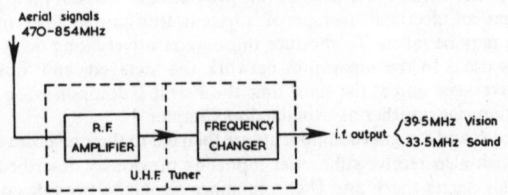

FIG. 8.1   THE TUNER STAGES

because noise generated in the early stages of the receiver may limit the useful gain. In addition, the r.f. stage must provide sufficient selectivity so that only the wanted channel signal is accepted. Any response to the 'image' signal must be low (about − 53 dB). The frequency-changer stage normally employs a self-oscillating mixer with the local oscillator tuned ABOVE the frequency of the aerial signal by the required receiver i.f. The oscillator voltage must not reach the aerial as aerial or direct radiation from the oscillator components may cause interference to other receivers. Therefore the receiver tuner must be adequately screened.

TRANSMISSION (LECHER) LINE TUNING

At Band 4 and 5 frequencies, conventional L, C tuning circuits are not practical owing to the very small values of inductance required. For example, a 5 pF capacitor requires an inductance of only approximately $0.012\,\mu$H to resonate at 800 MHz. This is the approximate self-inductance of a 0.7 mm diameter wire about 1 cm long. Fortunately, where ordinary L, C tuning is no longer feasible, it is possible to use tuned transmission lines.

Either an o/c $\lambda/2$ line or a s/c $\lambda/4$ line may be used. Fig. 8.2(a) shows a pair of transmission lines of length corresponding to $\lambda/4$ and short circuited at the far end.

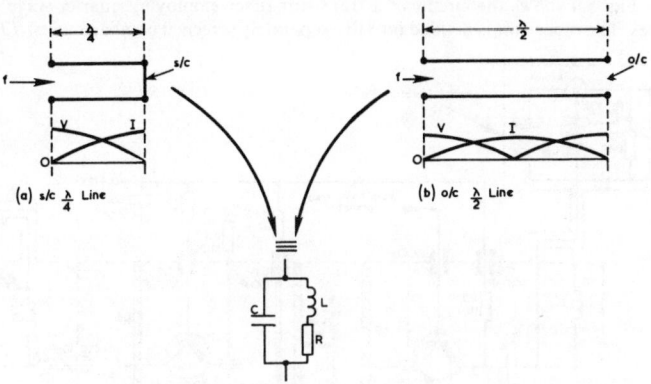

FIG. 8.2   USE OF TRANSMISSION LINES AS RESONANT ELEMENTS

Ideal voltage and current standing waves are as indicated (discussed in Chapter 6). If the frequency of the signal input to the line is raised above the frequency f to which the line is a quarter of a wavelength long, the line appears more than a quarter-wavelength long and offers a capacitive input impedance. At f where the line is exactly $\lambda/4$ long, the input impedance is a high resistance. At a lower frequency, the line appears less than a quarter of a wavelength long and the input impedance is inductive. These changes in impedance are exactly the same as those that occur in a parallel resonant circuit as the frequency is varied either side of the resonant frequency. Thus a shorted $\lambda/4$ line may be used to replace a parallel tuned circuit. An o/c $\lambda/2$ line [Fig. 8.2(b)] behaves similarly when the frequency of the signal input is varied about f to which the line appears a half-wavelength long.

The lines must be tunable to deal with signals in the range of 470–854 MHz. It is not practical to vary the physical length of the lines; their electrical length, or resonant frequency, may be varied by a variable capacitor placed at one end of the line as in Fig. 8.3. Increasing the value of the tuning capacitance has the same effect as increasing the physical length, i.e. the frequency of operation is reduced. Trimming capacitors are also included to 'track' the lines. The additional capacitance connected across the lines, of course, upsets the ideal voltage and current distribution given in Fig. 8.2. In the practical arrangement, one of the lines is formed by the chassis of the tuner (called the

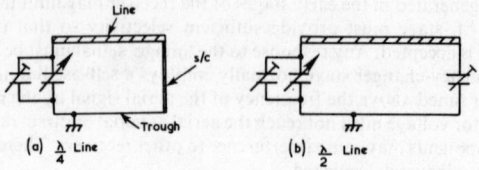

FIG. 8.3   TUNING THE LINES

'trough'). Coupling to or from a tuned line may be *via* a small loop of wire running parallel to the line or by connection to a suitable tap on the line.

Early valve u.h.f. tuners used $\lambda/2$ lines; the $\lambda/4$ lines could not be applied to valves because long leads were needed to connect to the electrodes of the valve. With the transistor u.h.f. tuner, which provides greater gain and a better noise performance than its valve counterpart, $\lambda/4$ lines were preferred both from the electrical and mechanical viewpoints. Some recent designs using 'electronic tuning' feature $\lambda/2$ lines.

Practical Tuner

Fig. 8.4 shows the circuit of a transistor tuner employing quarter-wave shorted lines. The tuner unit is divided into five separately screened compartments. TR_1 is the

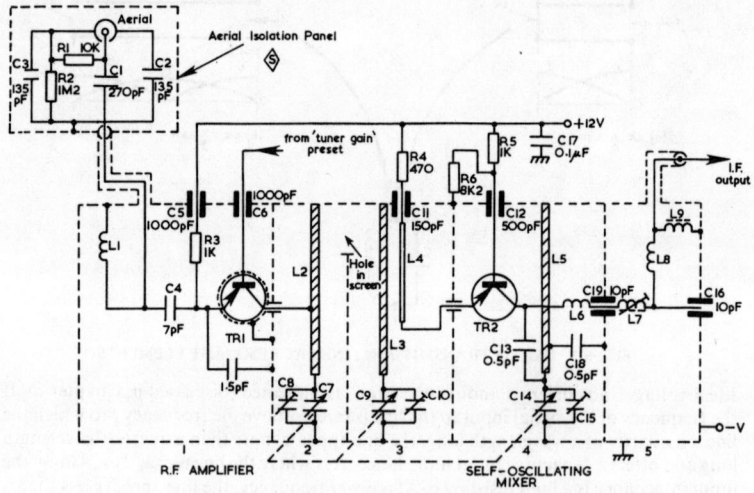

FIG. 8.4   U.H.F. TUNER UNIT USING GANGED TUNING CAPACITORS

r.f. amplifier stage and TR_2 serves as a self-oscillating mixer. Both transistors are connected in common base with C_6 grounding and TR_1 base and C_{12} grounding TR_2 base.

U.H.F. signals from the 75 Ω aerial are passed through C_1 *via* a length of coaxial cable and are developed across a small inductor L_1. This inductor acts as a low frequency shunt and thereby removes unwanted low frequency interfering signals. The signals are then coupled to the emitter of TR_1 through C_4. R_3 is the emitter stabilising resistor, and forward bias is applied to the base from a preset control on the main receiver panel. This control (local-distant control) may be set to adjust TR_1 gain to suit local reception conditions, *i.e.* the gain of TR_1 would be reduced if exceptionally strong signals overloaded the receiver. In designing the input circuit of TR_1 to match the 75 Ω aerial, the noise performance of TR_1 must be considered. The conditions of minimum noise and best matching tend to conflict; a small mismatch and loss of gain may be necessary to keep the noise at a low level. This is important because noise generated in TR_1 stage is amplified together with the signal.

Amplified signals are developed across the tuned line L_2 and to maintain the Q of the line, the collector is connected to the line *via* a suitable tap. The resonant lines L_2 and L_3 form a band-pass coupled circuit with energy transferred between the lines *via* a hole in the screening wall. These lines are tuned by C_7 and C_9 and provide the selectivity of the r.f. stage. Preset capacitors C_8 and C_{10} are used for trimming the operational range of the lines.

Signal energy from L_3 is coupled to the emitter of the mixer transistor TR_2 *via* a coupling loop L_4, running parallel to the tuning line. R_4 is the emitter stabilising resistor and base bias is obtained from the potential divider R_5, R_6. L_5 is the oscillator line, tuned by C_{14} and trimmed by C_{15}. Oscillatory energy is fed to the line from TR_2 collector *via* C_{13}. L_5 together with the inter-electrode capacitances of TR_2 form a Colpitts oscillator with feedback to across the emitter-base impedance. An accurate representation of the feedback circuit is difficult to show owing to the lead inductances and stray capacitances which become so important at u.h.f. As signal and oscillatory voltages are applied between base and emitter of TR_2, additive mixing takes place and the difference frequency (39·5 MHz vision and 33·5 MHz sound) is developed across the i.f. coil L_7 in the collector circuit. L_7 is tuned by C_{19} and C_{16}. The d.c. return for TR_2 collector is *via* the i.f. choke L_9 across which the i.f. output is developed. L_8 prevents oscillator radiation from the tuner and L_6 is an oscillator filter choke. The i.f. output is coupled to the i.f. amplifier stages by means of a low impedance coaxial cable link.

The tuning capacitors C_7, C_9 and C_{14} which are ganged together may be continuously varied over the u.h.f. bands or, as is more common, stations may be selected by a push-button mechanism which rotates the capacitors to preset tuning positions.

Capacitors C_5, C_6, C_{12} and C_{16}, etc. are of the 'lead-through' type. Fig. 8.5(a) shows the basic construction where the 'lead-through' wire forms one 'plate' of the

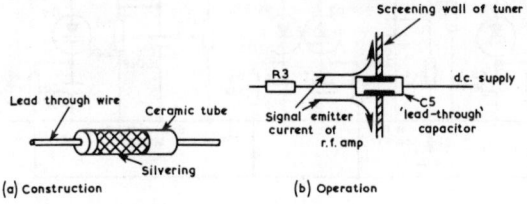

FIG. 8.5   'LEAD-THROUGH' CAPACITOR

capacitor and the silvering on the ceramic dielectric tube forms the other plate. The special feature of the capacitor is its low self-inductance, an essential characteristic in a capacitor to be used for decoupling at high frequencies. Diagram (b) shows a lead-through capacitor (C_5 of Fig. 8.4) fitted in the tuner screening wall and carrying d.c. to R_3. U.H.F. signal currents flowing in TR_1 emitter lead are decoupled by C_5 so confining them to the walls of the screening can. Similarly, any unwanted high frequency signal components of current present in the d.c. feed will not penetrate the walls of the screening can.

Aerial Isolation Panel

Receivers having one side of the mains supply (the neutral line) connected to the chassis must have means of protecting the user against an electric shock *via* the aerial socket should the chassis become 'live'. Fig. 8.4 shows an aerial isolation panel which provides protection. The aerial socket is isolated from chassis by C_1, C_2 and C_3 of such values that provide a low reactance at signal frequency but a high reactance at 50 Hz mains frequency. In addition, the voltage rating must be appropriate to the mains voltage. R_1 and R_2 provide a discharge path for static voltage built up on the aerial. These components must conform to safety standards (see symbols below) and must be replaced only by parts recommended by the manufacturers. An aerial isolation panel is absolutely essential when the receiver chassis is at half-mains voltage whichever way round the mains are connected to the receiver.

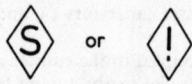

Electronic Tuning

In place of variable capacitors, varactor (vari-cap) diodes may be used to tune the resonant lines, a technique commonly referred to as 'electronic tuning'. A varactor diode has a capacitance (the depletion capacitance) that varies with the reverse voltage applied to it; capacitance is decreased as the reverse voltage is increased. A basic circuit of a u.h.f. tuner using vari-cap diodes is given in Fig. 8.6. The circuit has been simplified by not showing some of the components in their correct physical position and by omitting some of the decoupling capacitors, but the principle of operation remains the same.

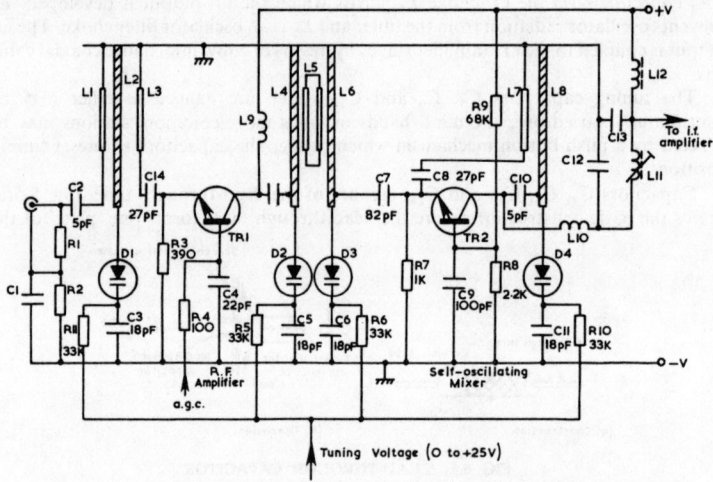

FIG. 8.6   ELECTRONIC TUNING USING VARACTOR DIODES

The tuner contains four short-circuited quarter-wave lines L_2, L_4, L_6 and L_8 with the varactor diodes D_1—D_4 used for tuning the lines. Signals from the aerial socket are fed to across the input wire loop L_1 via the isolating capacitor C_2. L_1 couples the signal to the first tuning line L_2 which is tuned by D_1. Signal energy is transferred from L_2 to TR_1 emitter via another coupling loop L_3 and C_{14} (d.c. block). R_3 is the emitter stabilising resistor and C_4 grounds the base of TR_1 to signals. Forward a.g.c. is applied to TR_1 base which reduces the gain of TR_1 when very strong signals are received.

TR_1 collector feeds the primary of a bandpass coupled line arrangement consisting of L_4 tuned by D_2 and L_6 tuned by D_3. Mutual coupling between the lines is by means of the wire coupling loop L_5. Signals to TR_2 emitter are fed from a tap on L_6 via C_7 (d.c. block). R_7 is the emitter stabilising resistor for TR_2, and the potential divider R_8, R_9 sets the base bias. C_9 grounds TR_2 base to signals. L_8 is the oscillator line and is tuned by D_4. Oscillatory energy is fed to the line from TR_2 collector via C_{10}. Feedback to sustain oscillations is via the coupling wire loop L_7 and C_8 to TR_2 emitter. The i.f.

resulting from the additive mixing action is developed across the i.f. coil L_{11} and fed out to the i.f. amplifiers. L_{12} provides the d.c. return from TR_2 collector and L_{10} is an oscillator filter choke.

The tuning voltage is supplied to the varactor diodes from a common supply *via filters* R_{11}/C_3, R_5/C_5, R_6/C_6 and R_{10}/C_{11}. These filters prevent coupling between the various sections; filter chokes may be used for this purpose. Since the positive tuning voltage is applied to the cathodes of the varactor diodes, they are placed in the required reverse bias state.

An example of the tuning voltage supply when ordinary push-buttons are used for changing channels is given in Fig. 8.7. The supply must be well stabilised or changes in the tuning voltage will occur with subsequent detuning of the resonant lines. In Fig. 8.7. the tuning voltage is derived from a 200 V d.c. supply using a zener diode to

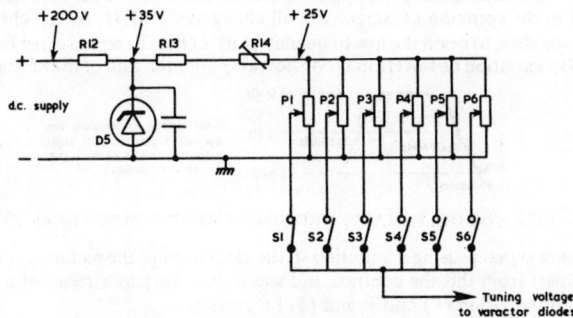

FIG. 8.7   TUNING VOLTAGE SUPPLY SHOWING TUNING POTENTIOMETERS (P_1–P_6)

provide a basic 35 V stabilised supply. In place of D_5 an integrated circuit stabiliser may be used. P_1—P_6 are the tuning potentiometers and the voltage across each is set to 25 V by R_{14} which determines the tuning range. By pressing any of six selector buttons which actuate the switches S_1—S_6, a tuning voltage may be supplied to the tuner from one of the potentiometers. It is therefore possible with this arrangement to select any of six preset channels.

Fig. 8.8 shows the relationship between the magnitude of the tuning voltage and the u.h.f. channels. The slope of the tuning which is dependent upon the varactor diodes is about 22 MHz per volt (Channels 21—50) reducing to about 10 MHz per volt (Channels 50—68). Thus a tuning voltage variation of, say, 100 mV will cause a frequency variation of between 1 MHz and 2·2 MHz depending upon the channel in use.

An advantage of electronic tuning is that the tuning potentiometers (which carry

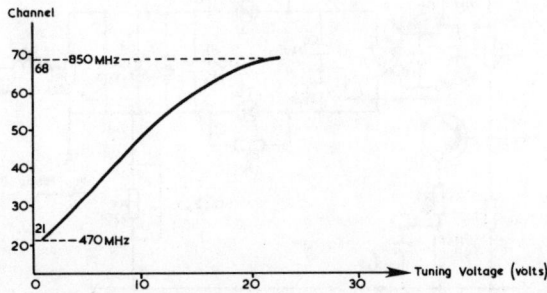

FIG. 8.8   RELATIONSHIP BETWEEN TUNING VOLTAGE AND U.H.F. CHANNEL FOR AN ELECTRONIC TUNER

only d.c.) and selector buttons may be conveniently mounted at the front of the receiver cabinet allowing the tuner itself to be mounted close to the i.f. panel.

Automatic Frequency Control (A.F.C.)

Automatic frequency control (sometimes called 'automatic fine tuning') is often used in colour receivers whether the tuner is mechanically or electronically tuned. A.F.C. is used to offset frequency drift in the local oscillator due to temperature or supply voltage variations and to correct errors in tuning caused by the mechanical tolerances (in variable capacitor tuners) or electrical tolerances in the varactor diodes and their supply circuits (in electronic tuners).

With the local oscillator working at, say, 800 MHz it only requires a 0·00125% drift in frequency to produce a frequency error of 1 MHz. Since the i.f. produced varies in sympathy with the oscillator frequency, the vision, sound and chrominance i.f.s presented to the common i.f. stages will all change by 1 MHz. As the chrominance signal (C) sits close to or on the low frequency 'skirt' of the i.f. response (see Fig. 8.9 and Chapter 9) a variation of 1 MHz may considerably alter the gain of the i.f. stages to the

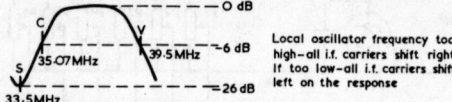

FIG. 8.9   RESPONSE OF I.F. STAGES (SHOWING CORRECT POSITION OF I.F. CARRIERS)

chrominance signal causing a variation in the saturation of the picture or even loss of colour. Apart from this the contrast and sound may vary as a result of a change in response to the vision (V) and sound (S) i.f. carriers.

Changes in the local oscillator frequency may be detected by measuring the frequency error of the vision i.f. by means of a frequency sensitive circuit. A common circuit used is the Foster-Seeley discriminator and a typical arrangement is given in Fig. 8.10 which will be described. The base of TR_1 is fed with a large amplitude vision i.f. signal supplied from a late stage in the i.f. amplifier or from the vision detector (when an integrated circuit is used). The drive should be sufficiently large to cause TR_1 to bottom so that its output is independent of the magnitude of the i.f. input. The amplified output of TR_1 is developed across L_1, C_5 (tuned to 39·5 MHz) and then fed via the secondary circuit L_2, C_6, C_7 (also tuned to 39·5 MHz) to the discriminator diodes D_1 and D_2. A reference signal is supplied from TR_1 collector to the centre connection of C_6 and C_7 which provide an artificial centre-tap on L_2. When the vision

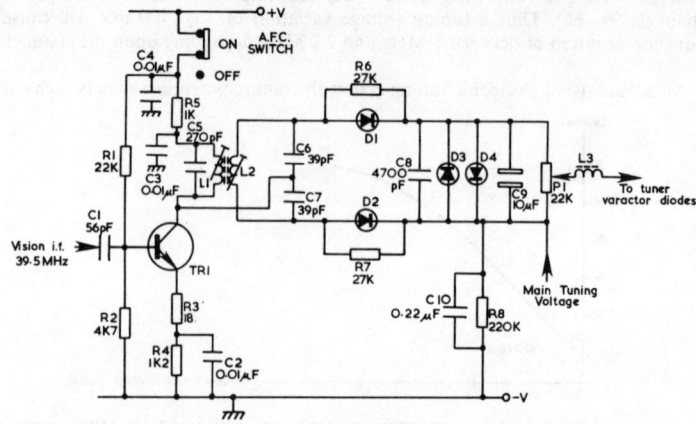

FIG. 8.10   A.F.C. DISCRIMINATOR

i.f. is correct D_1 and D_2 conduct by the same amount and C_6 and C_7 receive equal charges. As a result, the net d.c. voltage across the load P_1 is zero. Deviation of the vision i.f. from its correct frequency causes unbalance in the conduction of the discriminator diodes and unequal charges in C_6 and C_7. Thus a net d.c. voltage appears across P_1 with polarity depending upon the direction of frequency error. The magnitude of the 'error voltage' generated by the circuit is limited by the clipping diodes D_3 and D_4 to their forward voltage drop. The error voltage is fed out to the varactor diodes of the u.h.f. tuner in series with the main tuning voltage developed across R_8. Thus the error voltage either adds to or subtracts from the tuning voltage, depending upon the polarity of voltage across P_1 and pulls the oscillator in the correct direction to reduce the initial frequency error. It should be noted that if the oscillator drifts from its correct setting, the a.f.c. cannot make the vision i.f. exactly correct as an error voltage is required to pull the oscillator in the correct direction to reduce the frequency error. P_1 sets the 'holding' range of the a.f.c. circuit, *i.e.* the frequency range over which the circuit will track frequency drift in the tuner.

When tuning the receiver to a station the a.f.c. switch is set to the OFF position which inhibits TR_1 and prevents the generation of an error voltage. Once the station has been tuned in approximately, the a.f.c. is set to the ON position whereupon the circuit will correct any initial tuning error and subsequent drift in the local oscillator.

Touch Tuning

Touch-tuning is a completely non-mechanical method of changing channels and is used only with tuners employing electronic tuning. Any of six (usually but may be eight) desired channels can be selected simply by placing a finger on a button; one example of the 'electronics' involved is shown in Fig. 8.11.

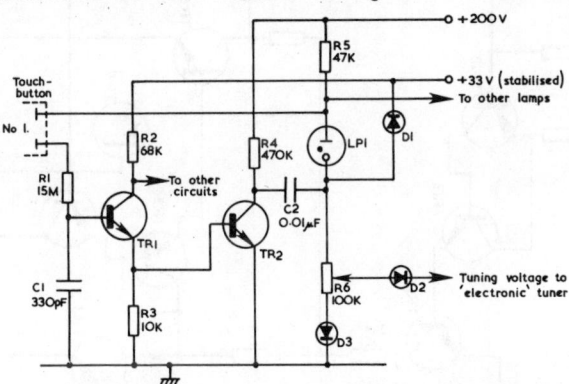

FIG. 8.11   TOUCH-BUTTON TUNING CIRCUIT (BUSH-MURPHY)

Six identical circuits are used (but only one shown here) with each circuit associated with a particular touch-button and neon indicator lamp. TR_1 and TR_2 act as switches but are normally in the OFF state. By bridging the contacts of the touch-button with the finger, base potential is applied to TR_1 *via* the high impedance circuit R_5, R_1 causing the transistor to turn ON. This action establishes a voltage across R_3 which causes TR_2 to conduct and its collector potential to bottom. The sudden fall in voltage at TR_2 collector is passed *via* C_2 to the neon lamp LP_1 causing it to strike. Current then flows in LP_1, R_6 and D_3 as the voltage at the junction of C_2, LP_1 rises positively. When the voltage exceeds 33 V, D_1 conducts clamping the potential at C_2, LP_1 junction to $+33$ V (stabilised). The conducting D_1 now holds the neon in the ON condition and it will remain so when the finger is removed from the touch-button. Thus $+33$ V (stabilised) is applied across the channel preset R_6 and D_3. R_6 feeds the tuner unit through the isolating diode D_2 which prevents current flow in R_6 from the varactor diode circuits in

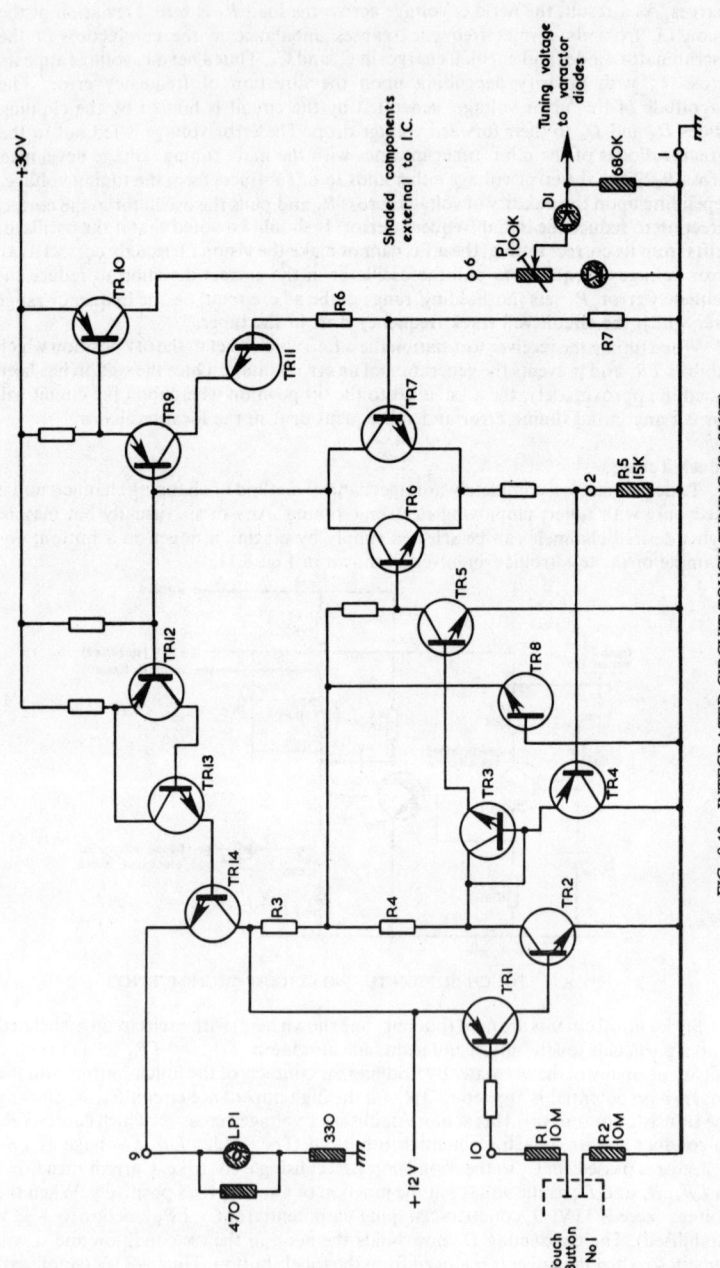

FIG. 8.12 INTEGRATED CIRCUIT TOUCH TUNING (SAS560T)

the tuner when another channel is selected. D_3 is used to compensate for the temperature coefficient of D_2.

When any other touch-button is bridged, the same action takes place in another circuit causing a different neon to strike and causing a different tuning potential to be applied to the varactor diodes in the tuner. However, because all lamps draw current through R_5 when they strike, operating another touch circuit will cause a voltage drop across R_5 thereby extinguishing LP_1 and removing the tuning potential from across R_6.

From the foregoing it will be seen that the following requirements are to be met by the touch-button tuning circuit: (a) fast operating switching action; (b) a latching device to maintain the circuit condition when the finger is removed; and (c) a form of reset facility. Integrated circuits are now being used for touch-tuning applications and one example used by I.T.T. is given in Fig. 8.12.

The diagram shows one touch-button circuit but there are four incorporated in the i.c. In the quiescent state, TR_5 is conducting due to the current flowing in R_3, R_4 and TR_3 (connected as a diode) from the $+12$ V supply. As TR_5 collector is at low potential, TR_6 is OFF thus TR_9, TR_{10}, TR_{11}, TR_7, TR_{12}, TR_{13} and TR_{14} are also OFF. When the contacts of touch-button No. 1 are bridged, TR_1 switches ON and this action switches ON TR_2. The fall in voltage at TR_2 collector switches OFF TR_5 causing the collector voltage of TR_5 to rise sharply. The rise in voltage is limited by TR_8 (connected as a zener diode) but is sufficient to cause TR_6 to come ON. This action results in TR_9 and TR_{10} coming ON. With TR_{10} conducting, $+30$ V is available at pin 6 of the i.c. which is fed to the varactor diodes in the tuner from P_1 via the isolating diode D_1. P_1 varies the d.c. voltage fed to the tuner and hence determines the tuning of the selected channel.

When TR_6 switches ON so also does TR_{12}, TR_{13} and TR_{14}, operating the lamp LP_1 from the $+12$ V supply. This lamp serves as a channel indicator.

Due to the current in R_6 and R_7 when TR_9 switches ON, TR_7 is made to conduct and this action maintains the circuit condition when the finger is removed from the touch-button (i.e. the circuit latches ON). The external resistor R_5 is common to the other touch-button circuits, so that when another touch-button is operated the voltage at pin 2 rises causing TR_7 to switch OFF thus resetting the circuit.

The four circuits in the i.c. are identical to Fig. 8.12 except that TR_4 is only required in one circuit so that when the receiver is first switched ON the No. 1 channel is selected. When the receiver is first switched ON there is no current in R_5 and TR_4 base is at chassis potential. Thus TR_4 is conducting via R_4 and R_3 from the $+12$ V supply. In consequence, TR_5 base is at low potential and the transistor is OFF thereby making touch-button circuit No. 1 active.

To provide eight channels two integrated circuits are required and the corresponding i.c. to SAS560T is the SAS570T. The latter i.c. is identical to the former except that the TR_4 stage of the channel 1 circuit (shown in Fig. 8.12) is not needed.

Remote Control of Channel Selection

Some receivers are fitted with remote control facilities. At one time the remote control unit was connected to the receiver by a multi-core cable, but ultrasonic systems are now more popular since no trailing lead is required. An ultrasonic wave is similar in nature to a sound wave, but the variations in air pressure occur at higher frequencies (about 28—45 kHz).

The basic idea of one system is illustrated in Fig. 8.13 and shows the operation during remote control of channel selection. The ultrasonic transmitter is contained in a hand unit and is powered by a 9 V battery. An L, C oscillator 1 (usually a Hartley or tuned collector) generates an electrical signal at one of a number of fixed frequencies, lying in the range 28—45 kHz. The oscillator output which is quite large (about 100 V peak) is fed to a capacitive type transducer 3 which generates the ultrasonic wave. The principle of operation of the transducer is similar to that of a capacitor microphone (but in reverse). A pair of plates, one fixed and one free to move, are separated by a dielectric as shown in Fig. 8.14. When an a.c. voltage is applied across the plates, the

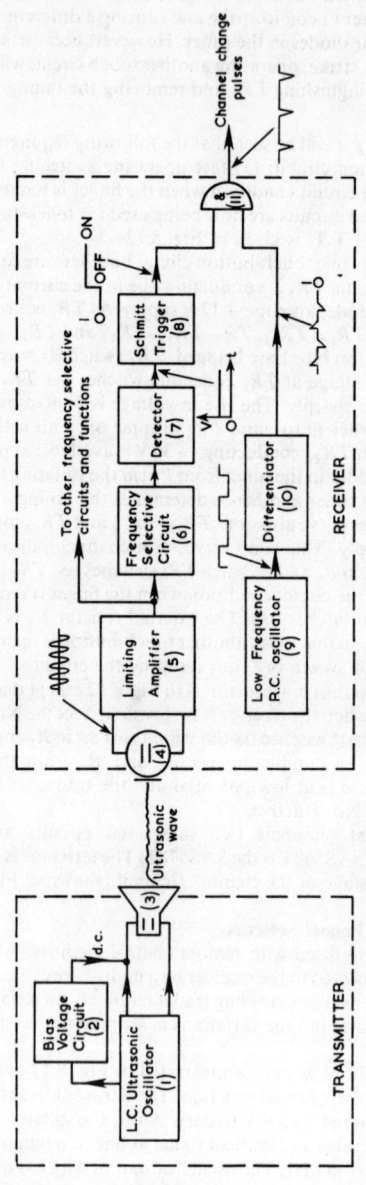

FIG. 8.13 BASIC IDEA OF ONE TYPE OF REMOTE CONTROL SYSTEM FOR CHANGING CHANNELS

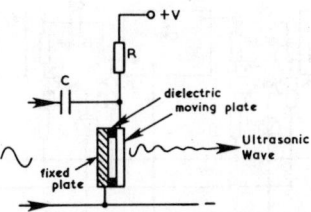

FIG. 8.14 CAPACITIVE TRANSDUCER

electrostatic forces between the plates vary causing the thin movable plate to vibrate thus producing an air wave. To ensure that the movable plate moves back and forth from a mean position a d.c. polarising voltage of about 80 V must be applied to the transducer (*via* the high resistor *R*). To obtain the d.c. polarising voltage, the oscillator output is rectified in the bias voltage circuit 2.

The advantage of using ultrasonics as opposed to an r.f. system is that the ultrasonic waves will not penetrate the walls of a house and cause interference to neighbouring receivers.

The number of functions performed by the remote control transmitter can vary. In simple systems only three functions are used but in others 15 or more may be provided. For example, the remote functions may be (i) Channel Selection; (ii) Volume Control; (iii) Brightness Control; (iv) Colour Saturation Control, etc. To control the various functions, selector buttons are provided on the hand unit. When a button is activated, the frequency of the *L,C* oscillator is changed (usually by bringing into circuit a different value capacitor). Thus each function is represented by a different ultrasonic frequency. It is important that the second and third harmonics of the line frequency (31·25 kHz and 46·875 kHz) are not used, or the receiver itself may cause false operation of the remote control. One or several buttons may be used for channel selection depending upon the system. It will be assumed here that only one button is used and when operated an ultrasonic wave of constant frequency is transmitted.

Turning now to the operation of the receiver, the ultrasonic wave from the transmitter is picked up by the transducer (4). As with the transmitter, a capacitive type transducer is employed the principle of operation being the same as that of the capacitor microphone. The electrical signal ouput of the transducer is fed to a high gain linear amplifier (5). This amplifier causes the signal to limit or 'square' which is necessary to provide a constant input to the Schmitt trigger (8), as the distance between the transmitter and receiver may vary. The output of the limiting amplifier is fed to a frequency selective circuit (6) (a tuned circuit) which selects the channel function frequency and passes the signal to the detector (7). In this stage the signal is detected and the resultant d.c is fed to the Schmitt trigger (8). The d.c applied to the voltage level sensitive Schmitt trigger causes it to go from the OFF state (0) to the ON state (1) at its output. It will remain in this state so long as the channel function button on the transmitter hand unit is depressed.

Block (9) is a free-running low frequency oscillator generating square waves at about 1 per second. The output from the oscillator is fed to a differentiating circuit (10) which feeds pulses as shown to the other input of the AND gate. When the logic levels 1 are present at the gate inputs, negative-going pulses occurring at a rate of 1 per second are available at the gate output. These pulses may be used to SEQUENTIALLY change channels in the tuner *via* a suitable circuit. Fig. 8.13 shows only the basic idea. Other gates and functions are required in a practical system and NAND gates are normally used. Many of the operations are performed by integrated circuits.

If the negative pulses available at the output of the AND gate of Fig. 8.13 are inverted, they may be used to sequentially switch the SAS560 integrated circuit, the operation of which was described using Fig. 8.12. For sequential operation of the i.c.

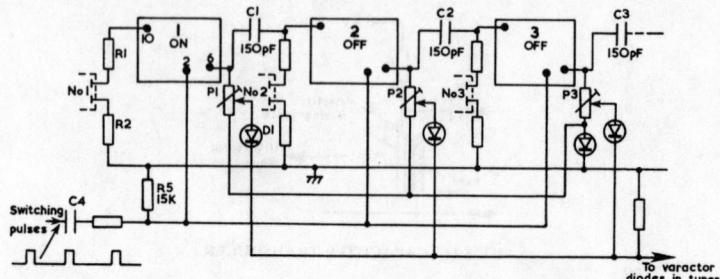

FIG. 8.15   SEQUENTIAL SWITCHING OF CHANNELS BY PULSES APPLIED TO I.C. (SAS506)

additional capacitors are required and a few of them (C_1—C_3) are shown in Fig. 8.15. Here three switching circuits are shown in block schematic form, with the output of one circuit capacitively coupled to the touch-button input of the following circuit (there are four switching circuits in the i.c.).

As previously explained, the No. 1 circuit becomes active when the receiver is switched ON and all other switching circuits are OFF. During remote operation when the first positive-going switching pulse is applied through C_4 to across R_5 (see Fig. 8.12), TR_7 will switch OFF due to the rise in its emitter voltage (the pulse will have no effect on the other switching circuits as they are OFF). This action will cause the voltage at pin 6 suddenly to fall. The fall in voltage is passed to the touch-button input of switching circuit No. 2 *via* C_1 causing the circuit to switch on and latch on as for normal touch-button operation. P_2 then supplies tuning voltage to the tuner changing the operation to channel 2. The next input pulse to across R_5 switches OFF switching circuit No. 2 which produces an output pulse to switch ON circuit No. 3 *via* C_2. P_3 then supplies tuning voltage to the tuner and channel 3 is selected. This action is repeated for each input pulse causing the sequential selection of the preset channels. The eighth input pulse switches OFF switching circuit No. 8 (using two i.c.s) which allows the No. 1 circuit to become active due to the operation of TR_4 (see Fig. 8.12). Releasing the channel button on the ultrasonic hand unit will prevent switching pulses reaching the channel selection circuit (due to the Schmitt trigger going OFF) thereby enabling the desired channel to be selected.

Another remote control system using pulse counting and dividing techniques is shown in Fig. 8.16 which illustrates the basic principle. This particular system provides up to 15 remote functions and each is initiated by operating an appropriate button on the transmitter hand unit. Channel selection is by four buttons which change the frequency of the transmitted ultrasonic signal.

At the receiver the incoming ultrasonic frequency is picked up by the capacitive transducer and the resulting electrical signal is fed to a limiting amplifier. The output of the amplifier is a square-wave type signal at the transmitted frequency (assumed to be 37·575 kHz) and this is fed to the Schmitt trigger. This circuit generates a square-wave of constant mark-space ratio with sharp rising and falling edges. The square-wave output of the Schmitt trigger is fed to a divider circuit (÷ 16) which is enabled by a suitable pulse for a period of 20 ms. During the enabling period an ultrasonic frequency of 37·575 kHz will give 751 input pulses to the divider, which thus provides 47 output pulses. These pulses are passed to a counter which provides an output in the form of a 4-bit binary digital code. For 47 input pulses, the counter completes two rotations, *i.e.* a count of 32 and on the third rotation reaches a count of 15 which corresponds to the binary code of 1111. At the end of the count, the enabling pulse to the ÷ 16 stage is removed, but the enabling pulse to the counter remains for a further 7 ms. During this period an enable pulse is applied to the latch. The latch ensures that the digital output of the counter which is changing during the 20 ms period is not passed to the decoder, thereby preventing the selection of random receiver functions. The binary digital output of the latch is then fed to a binary-to-decimal converter. With a binary input of

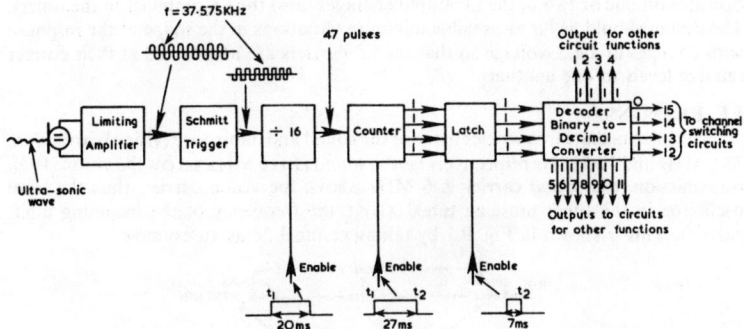

FIG. 8.16 BASIC IDEA OF ANOTHER TYPE OF REMOTE CONTROL SYSTEM USING PULSE
          COUNTING AND DIVIDING TECHNIQUES, SHOWING ULTRASONIC RECEIVER
          DURING CHANNEL CHANGE OPERATION

1111, the decimal 15 output goes to a low state (0) and this action selects channel 1 by operating a suitable channel selector circuit.

If a different channel selector button on the hand unit is operated the ultrasonic frequency is changed to, say, 36·775 kHz. The ÷ 16 stage will then give out 46 pulses during the enabling 20 ms period. The counter will go through two rotations and on the third rotation give a binary digital output of 1110. After being fed through the latch, the decoder decimal ouptut 14 will go to a low state (0) enabling the selection of channel 2.

Digital logic integrated circuits are used for many of the operations shown and other circuits are required to generate the enabling pulses. The use of the pulse counting and dividing technique eliminates the need for 15 separate channels in the receiver for the 15 transmitted ultrasonic frequencies.

CHAPTER 9

I.F. AMPLIFIER STAGES

The i.f. amplifying stages serve two principal functions:

(1) To increase the signal output of the u.h.f. tuner to an amplitude of several volts in order to drive the diode detector. For a nominal receiver sensitivity of 20 μV for 2 V across the vision detector load, a total gain between the aerial input and vision detector of about 100 dB is required. With a tuner gain of about 20 dB (an average value) an i.f. gain of 80 dB is necessary.

(2) To provide sufficient bandwidth to accommodate the vision and sound i.f. signals. Careful shaping of the i.f. response is required to correct for vestigial sideband working and to transmit the sound signal down the i.f. channel at a suitable level to create the intercarrier sound i.f. of 6 MHz. In addition, the response must give adequate rejection of unwanted signals, *e.g.* the adjacent channel vision and adjacent channel sound i.f. carriers.

The gain of the i.f. channel is not fixed but made automatically variable to deal with variations in the received signal strength. This is achieved by the use of a.g.c. which operates on one or two of the i.f. amplifier stages (also the r.f. amplifier in the tuner). The design should as far as possible minimise variations in the shape of the response with changes in a.g.c. voltage so that the i.f. carriers are maintained at their correct relative levels to one another.

I.F. RESPONSE

The intermediate frequencies used for the sound and luminance (vision) carriers are 33·5 MHz and 39·5 MHz respectively *i.e.* the sound i.f. is 6 MHz BELOW the vision i.f. On transmission, the sound carrier is 6 MHz ABOVE the vision carrier, thus the local oscillator in the tuner must be tuned ABOVE the frequency of the incoming u.h.f. carriers. This is shown in Fig. 9.1 by taking channel 26 as an example.

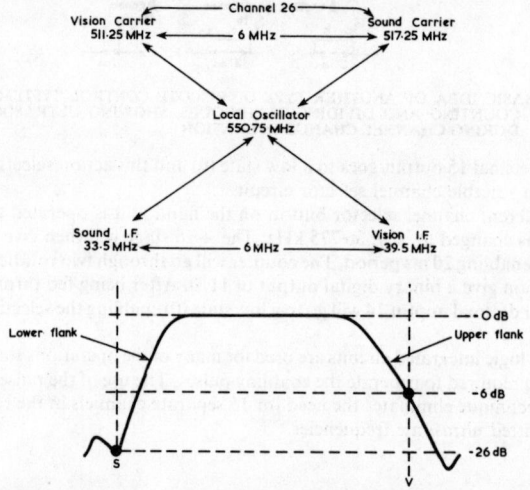

FIG. 9.1   RESPONSE OF THE I.F. STAGES TO THE VISION AND SOUND I.F. SIGNALS

A local oscillator frequency of 550·75 MHz will produce a difference-frequency of 39·5 MHz with a vision carrier of 511·25 MHz and a difference-frequency of 33·5 MHz with a sound carrier of 517·25 MHz. Thus the vision and sound i.f. carriers form a mirror image of the vision and sound u.h.f. carriers. The response must be so shaped that the vision i.f. of 39·5 MHz 'sits' 6 dB down from level response (0 dB). This is to correct for vestigial sideband working. The response to the sound i.f. of 33·5 MHz is depressed by about 26 dB and this is necessary to produce an optimum intercarrier sound i.f. signal.

The actual intermediate frequencies used are chosen to prevent interference to the displayed picture and reproduced sound. The frequencies are not critical and slight variations may be met with in practice, *e.g.* 33·4 MHz for sound and 39·4 MHz for vision.

The response of the i.f stages to the adjacent channel sound and adjacent channel vision carriers are shown in Fig. 9.2. This diagram illustrates how channels on either side of the channel to which the receiver is tuned (channel 26) can produce signals having frequencies which lie close to the bandpass of the i.f. channel. The sound carrier of channel 25 will give rise to a difference-frequency of 41·5 MHz with a local oscillator setting of 550·75 MHz (receiver tuned to channel 26) and the vision carrier of channel 27 will produce a difference-frequency of 31·5 MHz with this oscillator setting. Since these frequencies lie close to the i.f. bandpass they must be suppressed or interference may occur when a receiving aerial picks up the adjacent channel carriers from a transmitter different to the local one. Thus the response must be depressed at the Adj.V and Adj.S frequencies of 31·5 MHz and 41·5 MHz respectively. Note that these

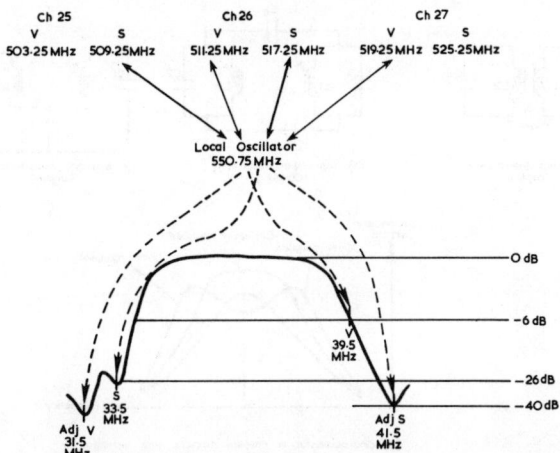

FIG. 9.2 TYPICAL I.F. RESPONSE SHOWING DEPRESSION OF ADJACENT CHANNEL VISION
AND SOUND I.F CARRIERS OF 41.5 MHz and 31.5 MHz

frequencies are separated by 10 MHz and lie 2 MHz above and below the wanted
intermediate frequencies of 39·5 MHz and 31·5 MHz.

Position of the Chrominance i.f.

In the video channel, the chrominance signal occupies a range of 4·43 MHz ±1
MHz. On transmission these frequencies are translated to 4·43 ±1MHz ABOVE the
vision u.h.f. carrier. Thus in the i.f. stages the chrominance signals are 4·43 ±1 MHz
BELOW the vision i.f. carrier, i.e. 39·5 MHz − 4·43 MHz = 35·07 MHz (±1 MHz). The
position of the chrominance i.f. on the response is usually at level response as in Fig.
9.3(a), or 6 dB down on level response as at (b). The response position of the
chrominance i.f. depends sometimes on the type of vision detector used and the aims of
the designer, to be discussed later.

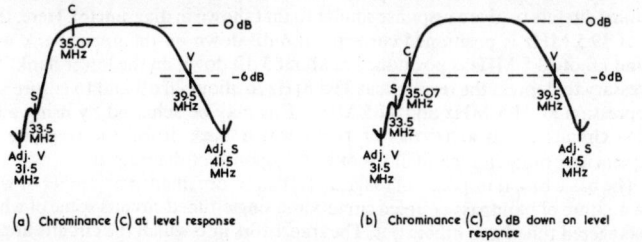

(a) Chrominance (C) at level response   (b) Chrominance (C) 6 dB down on level
                                                                                      response

FIG. 9.3 DIFFERENT POSITIONS FOR THE CHROMINACE I.F. (35.07 MHz) ON THE RESPONSE
CURVE

OBTAINING THE I.F. RESPONSE

Bandpass-coupled tuned circuits of the types shown in Fig. 9.4 are generally used to
achieve the desired i.f. response. In each of the arrangements shown in diagrams (a), (b)
and (c) L_1, C_1 and L_2, C_2 are tuned to the same frequency and are of the same inloaded
Q. By varying the degree of coupling, i.e. varying the mutual inductance in (a) or the
value of C_c in (b) and (c) any of the responses given in diagram (d) may be obtained.

To obtain sufficient overall bandwidth, three or four bandpass circuits may be
required which may be under-coupled, over-coupled or critically coupled. They may
all be tuned to the same frequency and damped with the aid of resistors to provide the

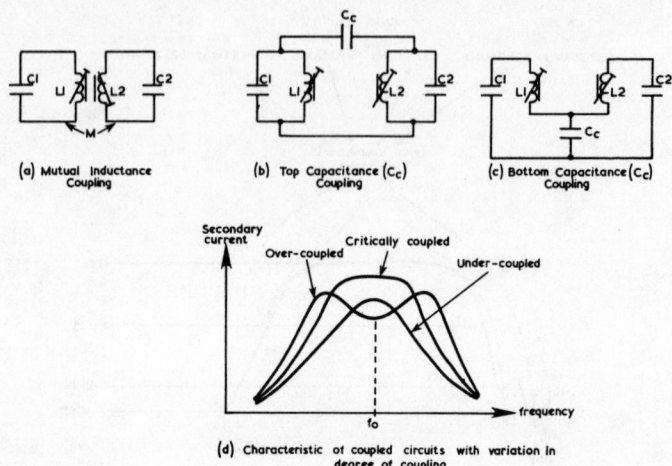

(a) Mutual Inductance Coupling

(b) Top Capacitance (C_c) Coupling

(c) Bottom Capacitance (C_c) Coupling

(d) Characteristic of coupled circuits with variation in degree of coupling

FIG. 9.4   BANDPASS COUPLED CIRCUITS

full i.f. bandwidth or alternatively, they may be staggered tuned, *i.e.* tuned to different frequencies. Three or four transistor amplifier stages will be required to obtain the necessary gain of about 80 dB, but an i.c. such as the MC1352 may be used to provide the gain.

Fig. 9.5 shows how an arrangement of four bandpass-coupled circuits connected between three discrete transistor amplifying stages may be used to obtain the required i.f. bandpass. It will be assumed that all the bandpass-coupled circuits are tuned to the same frequency of 36 MHz (about the centre of the i.f. bandpass). Bandpass circuits (1) and (3) are under-coupled and give a response similar to that shown in diagram (b). Over-coupling is used in circuits (2) and (4) resulting in a response such as that shown in diagram (c). The overall response of the four circuits are then as given in diagram (d). To increase the bandwidth, damping resistors may be connected across the bandpass-coupled circuits to give a response similar to that shown in diagram (e). Here, the vision i.f. of 39·5 MHz is positioned correctly at 6 dB down on the upper flank whilst the sound i.f. of 33·5 MHz is positioned at about 3 dB down on the lower flank. It is now necessary to depress the response at 33·5 MHz to about 26 dB and to ensure adequate suppression at 31·5 MHz and 41·5 MHz. This may be achieved by using additional tuned circuits called REJECTORS or TRAPS which 'suck down' the response at these frequencies producing the final overall i.f. response of diagram (f).

The basic broad response of diagram (e) can be obtained in a number of ways, *e.g.* by a mixture of bandpass-coupled circuits and single-tuned circuits some of which may be staggered tuned and others not. The transistors into which the circuits are coupled will normally provide some damping of the individual responses but additional resistor damping is nearly always used.

Rejectors or Traps

Examples of the type of rejector circuit to be met with in receiver circuits are given in Fig. 9.6.

L_1, C_1 in diagram (a) is a series-tuned trap and is mutually coupled to L_2 in the collector circuit of the i.f. amplifier. At the rejection frequency, to which L_1, C_1 is tuned, the circuit is of low impedance and absorbs energy from L_2, C_2, *i.e.* damps the collector circuit thus reducing the gain of the i.f. amplifier to the rejection frequency.

The parallel tuned rejector L_1, C_1 in diagram (b) is in series with the signal path. At the rejection frequency it offers a high impedance thereby reducing the amount of the

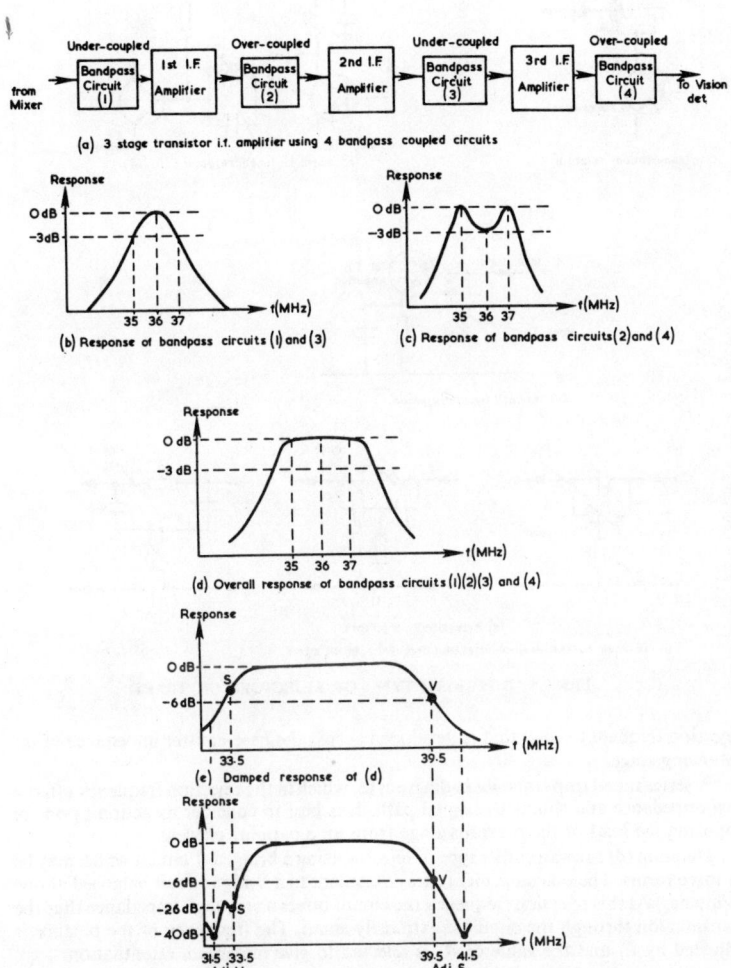

(a) 3 stage transistor i.f. amplifier using 4 bandpass coupled circuits

(b) Response of bandpass circuits (1) and (3)

(c) Response of bandpass circuits (2) and (4)

(d) Overall response of bandpass circuits (1)(2)(3) and (4)

(e) Damped response of (d)

(f) Final response after 'trapping'

FIG. 9.5   USE OF COUPLED CIRCUITS TO OBTAIN I.F. RESPONSE

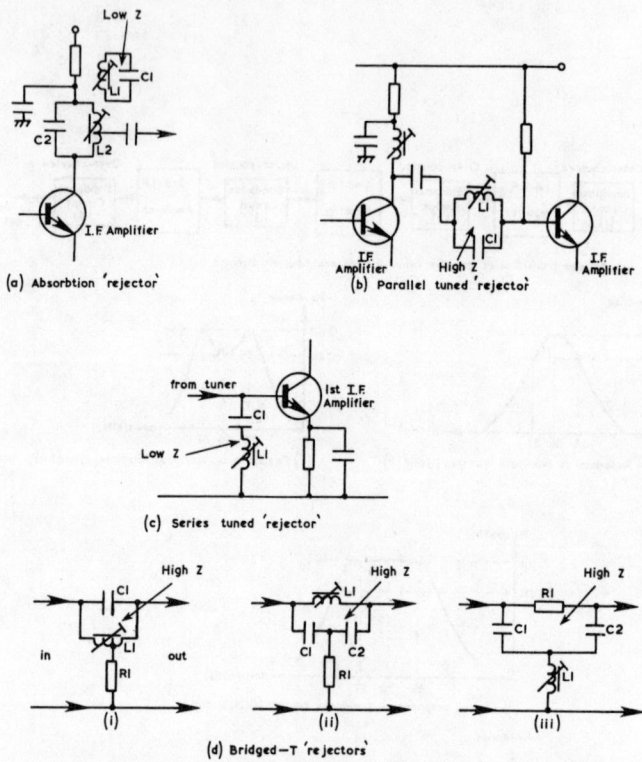

FIG. 9.6   DIFFERENT TYPES OF 'REJECTORS' OR 'TRAPS'

rejection frequency signal that is developed across the base-emitter impedance of the following stage.

A series tuned trap is shown in diagram (c) which at the rejection frequency offers a low impedance and shunts the signal path. It is best to consider its action as one of shunting the load of the previous stage from an a.c. point of view.

Diagram (d) shows another type of rejector using a bridged-T circuit which may be of three forms. These arrangements are equivalent to a bridge circuit balanced at one frequency. At this rejection frequency the circuit offers a very high impedance thus the transmission through the circuit is extremely small. The frequency of the balance is adjusted by L_1 and the value of R_1 is selected to give maximum attentuation.

I.F. AMPLIFIER CIRCUITS

An example of the i.f. stages of a monochrome receiver is shown in Fig. 9.7. Discrete components are used and the i.f amplifier is contained in four separate modules. Module 1 contains passive input selective filters and modules 2, 3 and 4 are active.

Module 1 contains a bridged-T rejector R_1, C_2, C_3 and L_1 which is tuned to 41·5 MHz. This rejector is responsible for the depression in the i.f. response at the frequency corresponding to the adjacent channel sound i.f. Also in this module are two series tuned rejectors C_4, L_2 and C_6, L_4 tuned to 33·5 MHz and 31·5 MHz respectively. These rejectors 'suck down' the response at the sound i.f and the adjacent channel vision i.f.

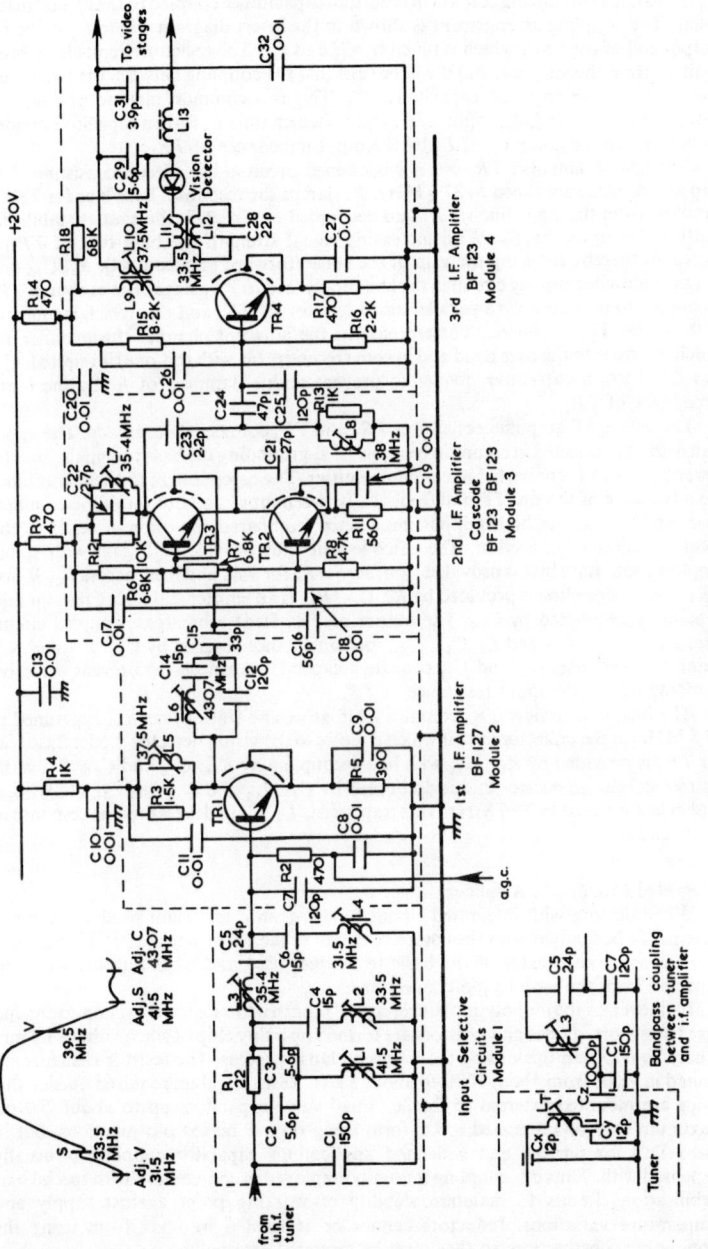

FIG. 9.7 I.F. STAGES (BUSH-MURPHY)

L_3 is a bandpass-coupling coil which is bottom capacitance coupled to the i.f. coil in the tuner. The coupling arrangement is shown in the insert diagram where L_T is the i.f. output coil of the tuner which is tuned by C_X and C_Y. The rejector circuits have been omitted from this diagram and it will be seen that the coupling between the two tuned circuits is *via* the common capacitance C_1. This is a common method of coupling between the tuner and i.f. amplifier. C_5 and C_7 which tune L_3 form a capacitive divider to prevent undue damping of L_3 by the input impedance of TR_1.

The first i.f. amplifier TR_1 uses a single-tuned circuit in its collector consisting of L_5 and self-capacitance tuned to 37·5 MHz. R_3 damps the response. Base bias for TR_1 is supplied from the a.g.c. line *via* R_2 and decoupled by C_8. R_5 is the emitter stabilising resistor decoupled by C_9. With increasing signal strength the base bias of TR_1 is increased thereby reducing the gain of the stage (forward gain control). R_4, C_{10} and C_{11} provide a decoupling circuit in the line supply feed to TR_1 stage. The output of TR_1 is coupled to module 3 *via* a parallel tuned rejector L_6, C_{14} and C_{12} which is tuned to 43·07 MHz. This frequency corresponds to the adjacent channel chrominance i.f. which lies close to the pass band and in some receivers (as with this one) is rejected. C_{15} and C_{16} form a capacitive divider to prevent undue damping of L_5 by the input impedance of TR_2.

The second i.f. amplifier consists of TR_2 and TR_3 connected in cascode. The upper transistor is connected in common base with C_{18} grounding the base to signals, and the lower transistor is connected in common emitter. The cascode arrangement combines the advantage of the small signal feedback between input and output of the common base circuit with the higher input impedance (compared to common base) of the common emitter connection. The cascode pair thus provide high gain with stable amplification. Base bias is provided from across R_8 for TR_2 and from across R_7, R_8 for TR_3. Line decoupling is provided by R_9, C_{17}. R_{11} is an emitter stabilising resistor and is suitably decoupled by C_{19}. The output of TR_3 feeds a bandpass-coupled circuit consisting of L_7, C_{22} and L_8, C_{24}, C_{25}, top capacitance coupled by C_{23}. R_{12} and R_{13} damp the response. C_{24} and C_{25} provide a capacitance divider to prevent excessive damping due to the input resistance of TR_4.

The final i.f. amplifier TR_4 contains a bifilar wound transformer L_9, L_{10} (tuned to 37·5 MHz) in the collector circuit which couples to the vision detector diode. Base bias for TR_4 is provided by R_{15}, R_{16} with line decoupling by R_{14}, C_{20} and C_{26}. R_{17} is the emitter stabilising resistor and is decoupled by C_{27}. L_{12} and C_{28} form an absorption type rejector tuned to 33·5 MHz. This trap assists L_2, C_4 in depressing the response at the sound i.f.

Integrated Circuit I.F. Amplifier

When dealing with integrated circuit functions in a television receiver, it is not essential to be familiar with the complete circuit of the particular i.c. but it is helpful if one has some knowledge of the basic techniques used and what outputs are to be expected from the various pin connections.

In linear i.c.s the monolithic (single crystal) construction is the most important and here transistors, diodes and resistors are formed on a tiny chip of silicon about 1·5 mm square by 0·25 mm thick using the diffused planar process. The resistor elements are limited in value from about 100Ω to about 5 k Ω. Resistor values required outside this range are mounted external to the i.c. Small value capacitors up to about 200 pF maximum may be fabricated in i.c. form using reverse biased p-n junctions, but in linear i.c.s for narrow and wideband applications capacitor coupling is usually dispensed with. Thus d.c. coupling is usually employed in conjunction with special bias stabilisation circuits to maintain stability of working point against supply and temperature variations. Inductors cannot be translated into i.c. form using the monolithic construction, so they must be mounted externally.

An example of an integrated circuit vision i.f. amplifier channel using a Motorola MC 1352 i.c. is shown in Fig. 9.8. All the response shaping is carried out external to the i.c. using the techniques previously outlined. Two pairs of bandpass-coupled circuits are

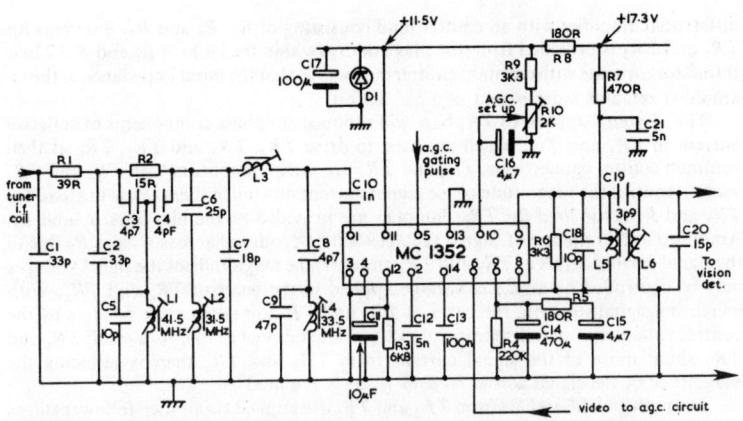

FIG. 9.8 INTEGRATED CIRCUIT I.F. AMPLIFIER (DECCA)

used to set the vision signal bandpass. One bandpass circuit is formed by the input coil L_3 which is bottom impedance coupled to the i.f. coil in the u.h.f. tuner. The other bandpass circuit is connected at the output of the i.c. and is formed by L_5, C_{18} and L_6, C_{20}. These tuned circuits are capacitively coupled by C_{19} and damped by R_6. The bridged-T network C_3, C_4, L_1 and R_2 provides rejection at the adjacent sound i.f. carrier of 41·5 MHz. Depression at the adjacent vision i.f. of 31·5 MHz is performed by the series tuned trap C_6, L_2. Another series tuned trap C_8, L_4 sucks down the response by the required amount at the sound i.f. of 33·5 MHz.

The i.c. performs two main functions: i.f signal amplifying and gated a.g.c. The i.f. signal amplifier provides a power gain of 53 dB and is divided into an a.g.c. controlled amplifying section and an i.f. output section. Fig. 9.9 shows the a.g.c. controlled amplifying section and illustrates some of the circuit techniques used in linear i.c.s.

The i.f. signal may be applied in push-pull to pins 1 and 2 or (as in Fig. 9.8) may be applied single-ended to either pin with the other pin grounded to signals. In Fig. 9.8 the i.f. input is applied to pin 1, and pin 2 is grounded by C_{12}. TR_1 and TR_2 form a

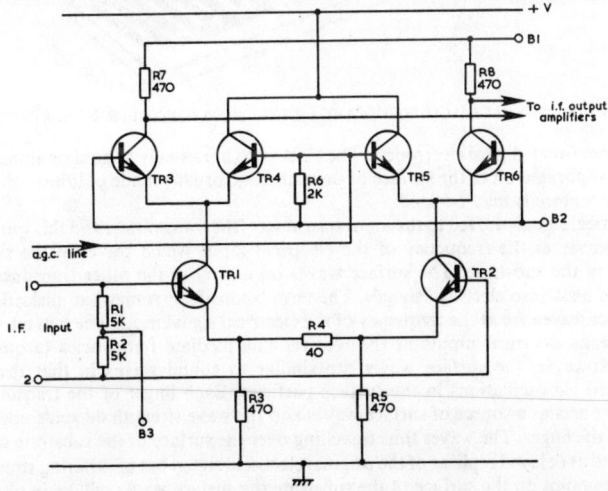

FIG. 9.9 PART OF I.F. AMPLIFIER OF MC1352 INTEGRATED CIRCUIT

differential amplifier with an emitter load consisting of R_3, R_4 and R_5. Base bias for TR_1 and TR_2 is provided from the bias line B_3 (within the i.c.) *via* R_1 and R_2. These transistors operate with constant emitter currents so that the input impedance of the i.f. amplifier remains independent of a.g.c. action.

The i.f. signal applied to TR_1 base will produce antiphase components of collector current in TR_1 and TR_2 which are used to drive TR_3, TR_4 and TR_5, TR_6 at their common emitter connections. TR_3 and TR_6 are the i.f. amplifiers with TR_4 and TR_5 used to control the magnitude of the signal current flowing in them. R_7 is the load for TR_3 and R_8 is the load for TR_6. Supplies are provided by the bias lines B_1 and B_2. Amplified and antiphase i.f. signal voltages will be produced across R_7 and R_8 due to the signal current drives at TR_3 and TR_6 emitters; the magnitude of the signal voltages may be controlled by an a.g.c. voltage applied to the bases of TR_4 and TR_5. With increasing signal strength, the bases of TR_4 and TR_5 are taken more positive by the control voltage causing these transistors to conduct more heavily. As a result TR_4 and TR_5 shunt more of the signal current from TR_3 and TR_6 thereby reducing the magnitude of the signal across R_7 and R_8 as is required.

The push-pull i.f. signals from TR_3 and TR_6 are applied *via* emitter-follower stages to the i.f. output amplifier connected as a differential pair. This pair provides a differential signal output of about 16 V peak-to-peak maximum between pins 7 and 8 of the i.c.

Surface Wave Filters

A recent development is the surface wave filter which may be used in place of the conventional LC bandpass and rejector tuning circuits in the i.f. channel.

A simplified representation of a surface acoustic wave filter (S.A.W.F.), is shown in Fig. 9.10. It consists of a piezo-electric substrate on which there are two transducers. Each transducer consists of two combed-shaped electrodes with the fingers enmeshing

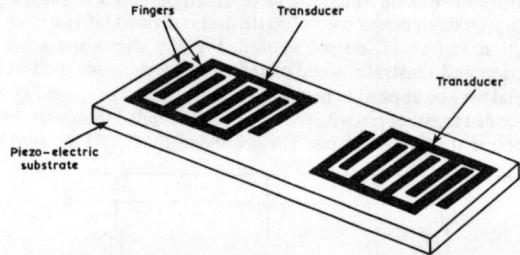

FIG. 9.10 SURFACE ACOUSTIC WAVE FILTER (SIMPLIFIED REPRESENTATION)

one another (inter-digital electrodes). The electrodes are usually of gold or aluminium and are evaporated on to the surface of the substrate, usually made of lithium niobate but other materials may be used.

Electrical signals are fed to the input transducer (the transmitter) and this generates surface waves at the frequency of the electrical input which travel in the surface particles of the substrate. The surface waves on arrival at the other transducer are converted back into electrical signals. The term 'acoustic' is somewhat misleading as the surface waves are at the frequency of the electrical signal input which in television terms means electrical inputs at the receiver intermediate frequencies (around 37 MHz). However, the surface waves are similar to sound waves in that they are propagated *via* oscillations in the surface particles. Each finger of the transmitting transducer acts as a source of surface waves and the wave strength depends upon the length of the finger. The waves thus travelling over the surface of the substrate can be considered as delayed replicas of the original electrical signal but of differing strengths. At a given point on the surface of the substrate the surface waves will be in phase to form a strong signal, whereas at other frequencies the waves will be out of phase and

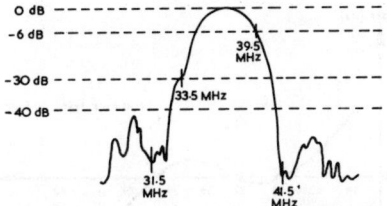

FIG. 9.11   AMPLITUDE RESPONSE OF TYPICAL SURFACE WAVE FILTER

thus cancel. By varying the form and dimensions of the transmitting and receiving transducer electrodes, bandpass characteristics may be obtained. Fig. 9.11 shows the amplitude-frequency response of a surface wave filter designed for television applications.

The velocity of the surface waves which penetrate the substrate to a depth of about 100 μm is of the order of about 1600 metres per second, depending upon the substrate material. At 37 MHz the wavelength of the surface wave is only about 40 μm, thus the dimensions of the filter will be in millimetres. The design of a filter is complex and one manufacturer (Mullard) uses a computer to design its filters. The performance of the filter not only depends upon the geometry of the electrodes but also on the properties of the substrate material. A substrate with a low temperature coefficient is chosen as the velocity of the surface waves varies with temperature. Also, the substrate material affects the 'insertion loss' of the filter (typically 17 dB). The filter must be encapsulated to protect the polished substrate from dirt and moisture.

As a circuit element measuring approximately $21 \times 14 \times 6$ mm the filter is fitted with four pins and may be soldered on to a printed circuit board panel. It may be used in monochrome or colour receivers and Fig. 9.12 shows its position in the i.f. amplifying chain. The filter may be driven from a preamplifier the input of which is the i.f. output

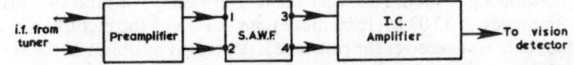

FIG. 9.12   USE OF S.A.W.F. ELEMENT IN MONOCHROME OR COLOUR RECEIVER

of the receiver tuner. The bandpass output of the filter is then fed to an i.c. which provides the receiver i.f. gain. It is claimed that a surface wave filter gives improved amplitude and group delay characteristics over conventional *LC* circuitry, as well as obviating critical adjustments during receiver production.

Group Delay

It has been seen that the amplitude response of the i.f. channel ensures that the various frequency components of the vision signal are correctly reproduced. The placing of the vision i.f. carrier 6 dB down on the upper flank of the response curve corrects for vestigial sideband working. Also, the relative levels of the other components including the chrominance signal assist in the correct reproduction of luminance and colour information on the c.r.t. screen.

Apart from amplitude response, the phase response of a receiver is also important. The various frequency components of the vision signal in passing through the restricted bandpass of the i.f. stages suffer differing delays, *i.e.* some components arrive relatively earlier and some relatively later than others. To avoid picture distortion due to these varying delays, the phase response of the receiver i.f. channel must be so designed that the delay of the sideband components is the same (or very nearly so). It is possible to measure the time delay of the sideband components in passing through the i.f channel relative to a particular frequency component and to plot these in the form of a 'group delay characteristic' as in Fig. 9.13

Ideally, the group delay response would be a straight line over the vision bandpass. The main cause of departure from the ideal is the phase shift produced by the high Q

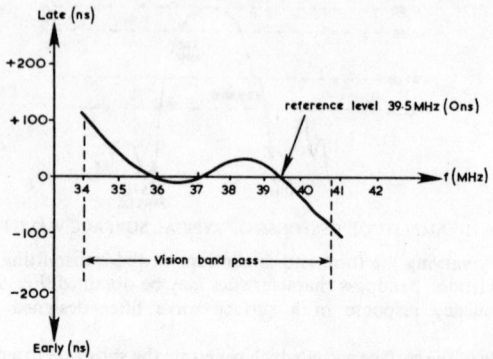

FIG. 9.13   GROUP DELAY CHARACTERISTIC OF TYPICAL VISION I.F. CHANNEL

rejector circuits at the edges of the vision bandpass. Surface wave filters give improved response with a group delay of the order of ± 40 nano seconds relative to 39·5 MHz (0 nano seconds).

The effect of group delay errors is to exaggerate overshoot effects and when signal components arrive early to cause preshoots or undershoots. Overshoot of the video signal causes white borders to the right of black vertical lines, and black borders after white vertical lines in the picture. Undershoot results in white edges to the left of black vertical lines and black edges to the left of white verticals.

Response of Mono and Colour Receivers

The i.f. response of the two types of receivers is the same apart from one or two restrictions placed on the response of the colour receiver. In a colour receiver it is important that the a.g.c. should not vary the response shape around the chrominance subcarrier frequency of 35·07 MHz by more than ± 1 dB as the picture saturation will alter. In a monochrome receiver the response is depressed by about 26 dB at the sound i.f. of 33·5 MHz to prevent the vision signal modulating the intercarrier sound i.f. of 6 MHz which causes 'buzz on sound'. In a colour receiver the rejection is increased to about 40 dB to reduce the level of the beat between the sound i.f. of 33·5 MHz and the chrominance subcarrier of 35·07 MHz which produces an obtrusive pattern of 1·57 MHz on vision. Less depression of the 33·5 MHz sound is required in a colour receiver when a synchronous vision detector is used.

In Chapter 5 it was mentioned that since the colour signal subcarrier and its sidebands all lie in a single sideband of the vestigially transmitted signal, distortion of the detected signal occurs when a diode envelope detector is used. This form of distortion, 'quadrature distortion', increases rapidly with the depth of modulation. The receiver i.f. response should not therefore be greater at the colour subcarrier of 35·07 MHz than at the vision i.f. of 39·5 MHz as otherwise the depth of modulation (as far as the diode detector is concerned) is effectively increased. Thus in many receiver designs the chrominance subcarrier of 35·07 MHz sits 6 dB down on the i.f. response, *i.e.* the same as the vision i.f. of 39·5 MHz.

A very satisfactory way of dealing with quadrature distortion is to use a separate i.f. amplifier and detector for the chrominance signal as shown in Fig. 9.14. By shaping the response of the chrominance i.f. amplifier with the aid of bandpass circuits, a response similar to that shown may be obtained. Here the response at the chrominance i.f. subcarrier of 35·07 MHz is made lower than that of the vision i.f. carrier of 39·5 MHz. This effectively reduces the depth of modulation and the distortion. After amplification and response shaping in the chrominance i.f. stage the signals are fed to a diode detector which delivers at its output chrominance signals of 4·43 MHz ± 1 MHz. There may be unwanted frequency components at the detector output, *e.g.* 6 MHz but these may be removed with the aid of rejector circuits. The use of two detectors and

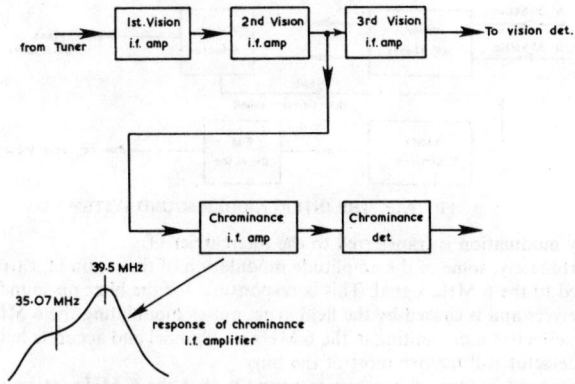

FIG. 9.14   USE OF SEPARATE CHROMINANCE I.F. AMPLIFIER AND DETECTOR

essentially two i.f. amplifiers with different frequency responses gives very good results. The use of a synchronous vision detector may also prevent quadrature distortion.

INTER-CARRIER SOUND

The reasons for its use in 625-line television are:

(1)   Because frequency modulation is used for television sound, the centre frequency presented to the f.m. demodulator must remain constant otherwise the sound will be distorted (see Fig. 9.15). If 33·5 MHz sound signals were fed to the f.m. demodulator severe distortion would result from drift in the u.h.f. tuner. For example, with a local oscillator working at 800 MHz a frequency drift of 125 kHz represents a drift in tuning of only

$$\frac{0 \cdot 125}{800} \times 100 \quad \% = 0 \cdot 015\%$$

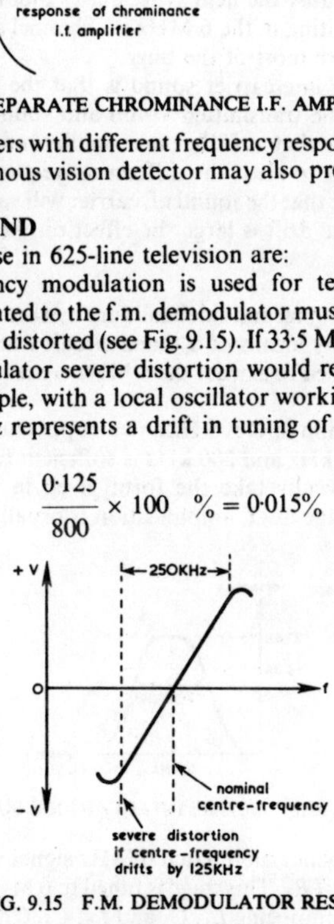

FIG. 9.15   F.M. DEMODULATOR RESPONSE

(2)   Also, there may be drift in the i.f. tuning which has similar results but not so serious as local oscillator drift.

In the intercarrier sound system (Fig. 9.16) both vision and sound i.f. carriers are passed through the vision i.f. stages to appear at the vision detector. As well as acting as an envelope detector for vision signals, the diode acts as a mixer to produce beats between the vision and sound i.f.s. The vision i.f. acts as a kind of 'local oscillator' signal and due to the non-linearity of the diode's characteristic additive mixing takes place. As a result a difference-frequency of 6 MHz is produced between the vision i.f. of 39·5 MHz and the sound i.f. of 33·5 MHz. The 33·5 MHz input to the detector is frequency modulated thus all its side-frequency components will beat with the vision i.f. to produce difference-frequencies on either side of the 6 MHz intercarrier, i.e. the

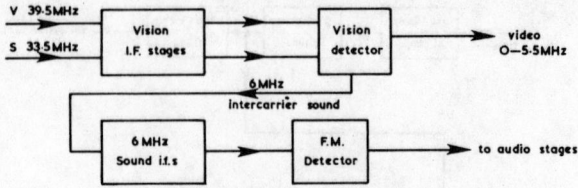

FIG. 9.16   THE INTERCARRIER SOUND SYSTEM

frequency modulation is transferred to the intercarrier i.f.

Unfortunately, some of the amplitude modulation of the vision i.f. carrier is also transferred to the 6 MHz signal. This is responsible for the buzz on sound heard in some receivers and is caused by the field sync. pulses modulating the 6 MHz signal. However, effective a.m. limiting in the 6 MHz i.f. channel and accurate balancing of the f.m. detector will remove most of the buzz.

The main advantage of intercarrier sound is that the 6 MHz intercarrier i.f. is accurately determined by the transmitted vision and sound carriers and not by the stability of the u.h.f. tuner oscillator. If the tuner oscillator drifts causing the frequency of the vision and sound i.f.s to alter, the difference-frequency remains constant at 6 MHz. The only effect will be that the sound i.f. carrier will vary in amplitude relative to the vision i.f. and unless the drift is large the effect on the sound will be negligible.

6 MHz I.F. Stages

The 6 MHz intercarrier signal and sidebands are extracted from the vision detector or one of the video stages by means of a tuned circuit which is resonant to 6 MHz. Examples of the methods used to extract the 6 MHz signal are considered in Chapter 10.

With a frequency deviation of ± 50 kHz and an upper audio frequency of 15 kHz a bandwidth of between 180 kHz and 200 kHz is sufficient for the 6 MHz i.f. channel, thus the response will typically take the form given in Fig. 9.17. When discrete transistors are used, one stage of i.f. amplification is usually provided for the 6 MHz signal as shown in Fig. 9.18.

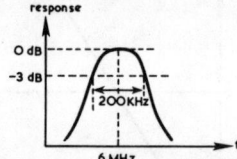

FIG. 9.17   RESPONSE OF INTERCARRIER SOUND I.F.S.

Here the extracted frequency modulated 6 MHz signal is fed *via* C_1 to the base of the i.f. amplifying transistor TR_1. This stage is tuned to 6 MHz in its collector circuit by the bandpass transformer comprising L_1, C_3 and L_2, C_4, C_5. In addition to amplifying the 6 MHz signal the stage also provides amplitude limiting to reduce the effects of 'intercarrier buzz'. The output from the secondary of the band pass transformer is transferred to D_1 and D_2 operating in a ratio detector circuit which demodulates the intercarrier signal (see Chapter 12).

Another example of a 6 MHz i.f. channel is given in Fig. 9.19 where a ceramic filter (CF_1) is used in place of the conventional LC tuning circuit and an i.c. is used to provide the gain. A ceramic filter makes use of the piezo-electric property of certain processed ceramic materials. The Q of such a device is much higher than can be obtained with ordinary LC circuits hence it gives good selectivity. Its construction is shown in Fig. 9.20. The ceramic material is given two coatings on one side as shown, and a continuous coating on the other side which is connected to the earthy side of the

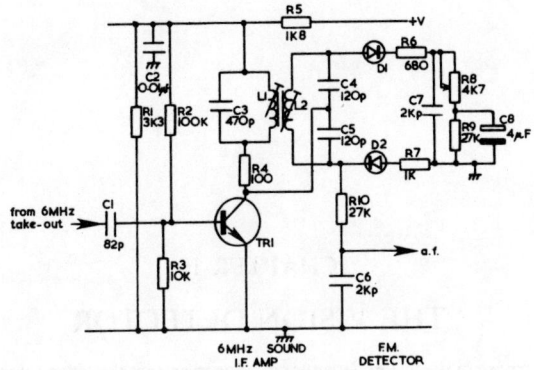

FIG. 9.18 INTERCARRIER I.F. AMPLIFIER

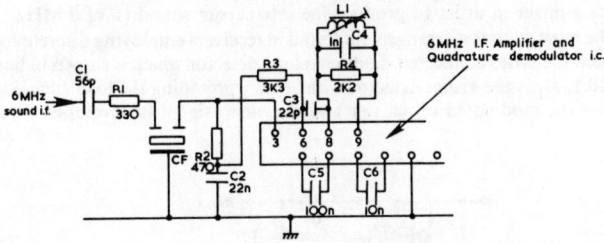

FIG. 9.19 INTERCARRIER I.F. AMPLIFIER USING CERAMIC FILTER AND I.C. AMPLIFIER

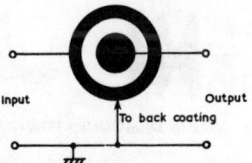

FIG. 9.20 CERAMIC FILTER

circuit. The input signal is fed to one coated area and the output taken from the other. The filter does not provide any gain but introduces some insertion loss (about 5 dB). The filters are not tunable and must be correctly matched at input and output (usually 330 ohms). In matching the filter to the required terminating impedance at input and output, account must be taken of the input and output impedances of the connecting circuits. Ceramic filters may be used in cascade to narrow the bandwidth but at the expense of an increase in insertion loss.

In Fig. 9.19 R_1 and R_2 are used for matching to the ceramic filter at input and output. The insertion loss of the ceramic filter is made up by the gain of the i.c. amplifier stages. Typically four or five differential amplifier stages are used, connected in cascade with d.c. coupling throughout. Resistance loads are used for the differential amplifier stages, of course, thus all the selectivity of the 6 MHz i.f. channel is provided by the ceramic filter. The high gain of the i.c. amplifiers causes limiting of the signal thus removing amplitude modulation of the i.f. envelope. Typically, an i.f. input to the i.c. of about $100 \mu V$ will cause limiting to occur. The i.c. may also contain a quadrature demodulator for detecting the f.m. signal. This will be discussed in Chapter 12.

CHAPTER 10

THE VISION DETECTOR

THE purpose of the vision detector circuit is to extract the video information from the i.f. carrier and to filter out the i.f. component before passing the video signal to the video stages. In an intercarrier sound receiver the vision detector also acts as a mixer in order to produce the intercarrier sound i.f. of 6 MHz.

The most common arrangement found in receivers employing discrete components has been the series connected diode envelope detector which is shown in basic form in Fig. 10.1. D_1 is the vision detector with R_1, C_1 providing the load time-constant. D_1 rectifies the modulated i.f. carrier and the video signal is developed across R_1. C_1

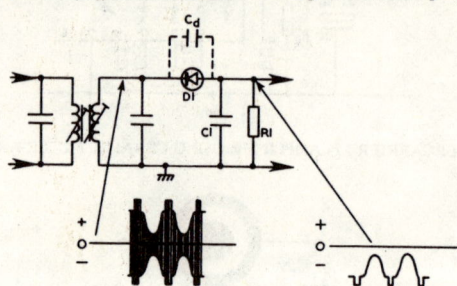

FIG. 10.1   VISION DEMODULATOR (BASIC CIRCUIT)

operates as a reservoir capacitor and charges up through the diode during its conducting period. When the diode is OFF (in this case during the positive-going half-cycles of the i.f. carrier) C_1 discharges through R_1. To obtain a linear output the diode should receive a modulated i.f. input of about 3 V to 5 V peak-to-peak so that operation is confined to the linear part $A—B$ of the diode's characteristic as shown in Fig. 10.2. Small input signals will have some non-linear distortion due to the lower bend of the diode's characteristic.

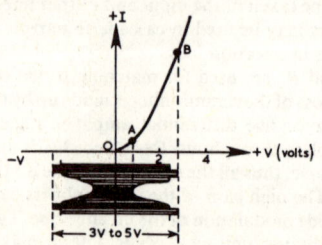

FIG. 10.2   OPERATION CONFINED TO LINEAR PORTION OF DIODE'S CHARACTERISTIC

124

Choice of C_1, R_1 Values

In arriving at suitable values for the detector time-constant the following factors have to be considered:

(a) The capacitance of C_1 should be large compared with the diode capacitance C_d. This is to prevent excessive i.f. signal appearing in the output when the diode is non-conducting due to the potential divider formed by C_1 and C_d across the input of the detector.

(b) The resistance of R_1 should be large compared with the forward resistance of the diode to provide high detector efficiency.

(c) The time-constant must be long compared with the periodic time of the i.f. carrier to reduce the amount of i.f. ripple in the output.

(d) The time-constant must be short compared with the periodic time of the highest modulating frequency (5·5 MHz) so that the output can follow the high frequency variations of the envelope.

(e) The detector load must provide a suitable impedance as seen by the final i.f. transformer to give correct tuning and damping.

In view of the above factors, some of which are conflicting, the design of the detector tends to be complex but the main aim is to provide a video output with little distortion by sacrificing some detector efficiency. To give an idea of suitable values used consider the following.

The periodic time of one cycle of the vision i.f. of 39·5 MHz

$$\frac{1}{39 \cdot 5 \times 10^6}\,\text{s} \simeq 0 \cdot 025 \ \mu\text{s}$$

and the periodic time of one cycle of the highest video frequency of 5·5 MHz

$$= \frac{1}{5 \cdot 5 \times 10^6} \simeq 0 \cdot 18 \ \mu\text{s}.$$

The time-constant R_1, C_1 should be greater than $0 \cdot 025$ μs but less than $0 \cdot 18$ μs. Suppose that a time of $0 \cdot 1$ μs is decided upon as a compromise value. If C_1 is made $20 \times C_d$ and C_d is about 1 pF (typically), $C_1 = 20$ pF.

$$\text{Therefore } R_1 = \frac{\text{time-constant}}{C_1}$$

$$= \frac{0 \cdot 1 \times 10^6}{20 \times 10^{-12}} = 5 \ \text{k}\Omega.$$

In a practical circuit the value of the physical C_1 may be less than suggested due to the stray capacitance across the circuit. In some cases C_1 is omitted altogether, the capacitance being formed entirely by the stray capacitance.

I.F. Filter

The purpose of the i.f. filter is to remove the residual i.f. ripple at the output of the detector and to do so without unduly attenuating the video signal. The ratio of vision i.f./highest video frequency is only about 7:1 and calls for efficient filtering. The RC filter used in radio receivers is not suitable. If the value of the series resitive element were high enough to attenuate the vision i.f. it would severely attenuate the video signal due to the low value of detector load resistance used. Thus the resitive element is replaced by an inductor as shown in Fig. 10.3.

The value of the series inductor is such that at the i.f. L_1 has a much higher reactance than that of C_2 but at video frequencies its reactance is lower that that of C_2.

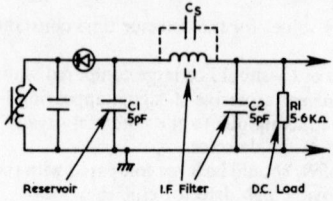

FIG. 10.3   I.F. FILTER

As a result, the i.f. is 'dropped' across L_1 and the video is developed across C_2/R_1. Sometimes it is arranged that L_1, together with its self-capacitance, form a parallel tuned circuit resonant to the i.f., thereby providing additional rejection of the i.f. at the detector output. Particular attention is given to effective filtering of the video signal because if the i.f. finds its way into the video amplifier it may cause i.f. instability as a result of reradiation. For this reason the vision detector and the filter circuit are usually housed in the final i.f. amplifier screening can. Sometimes a two-section LC filter is used to give more efficient filtering.

Polarity of Output

With a sound receiver it is unimportant which way round the diode is connected in the circuit since the output, being a.c., has no polarity. In a television receiver the polarity of output is most important since, if the signal is reversed, the white portion of the picture will appear black and *vice versa*.

The polarity of the diode depends upon the number of video stages and the method used to modulate the cathode-ray tube. Either grid or cathode modulation of the c.r.t. may be employed but with modern receivers cathode modulation is invariably used. Assuming that one video amplifier is fitted then the polarity required from the vision detector is as shown in Fig. 10.4. In cathode modulation a positive-going video signal from the detector results in a negative-going video signal at the cathode of the c.r.t. which is that required for cathode modulation of the c.r.t. beam current.

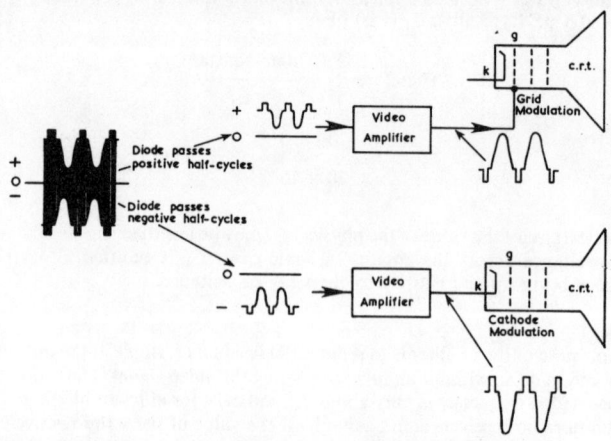

FIG. 10.4   GRID AND CATHODE MODULATION OF C.R.T.

D.C. Component

The vision detector, being a rectifier, produces a d.c. component in its output in addition to the video and i.f. components. For true picture fidelity the d.c. component

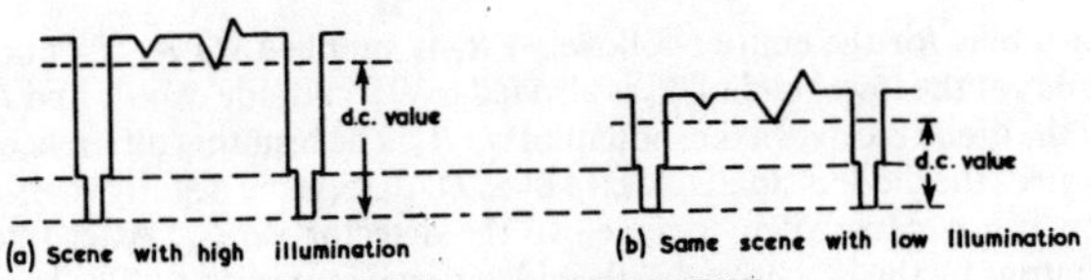

(a) Scene with high illumination   (b) Same scene with low illumination

FIG. 10.5   D.C. COMPONENT

should be retained as it represents the average scene brightness, see Fig. 10.5.

Diagram (a) shows one line of video information of a particular scene where the studio lighting is assumed to be of high intensity producing an average or d.c. value for the waveform as shown. If the studio illumination is made less intense to create, say, the impression of a change from daylight to dusk, the video signal will contain the same luminance variations but the d.c. component will be smaller as shown in diagram (b): Clearly, the d.c. component must be preserved if the impression of a change of scene illumination is to be conveyed to the viewer. To maintain the d.c. component, d.c. coupling should be used from the vision detector ouptut right up to the modulating electrode of the c.r.t., *i.e.* no coupling capacitors will be fitted.

Detector Emitter-Follower

The output of the vision detector feeds the video amplifier stage. When a common-emitter transistor video amplifier stage is used it is normally fed from the detector *via* an emitter-follower stage. This is to prevent the low input impedance of the video amplifier loading the vision detector. With an emitter load (R_L) of 500 Ω and an h_{fe} of 75, the input impedance of the common collector stage is approximately $h_{fe} \times R_L = 37.5$ kΩ. This value of impedance will have negligible shunting effect on a detector load resistance of 3—6 kΩ. The output impedance of such a common collector stage with an R_g of 5 kΩ (the detector load) across its input is approximately $R_g/h_{fe} = 66$ Ω in parallel with the 500 Ω emitter load, *i.e.* about 58 Ω. The low output impedance of the emitter-follower stage facilitates matching to the input of the video amplifier.

An example of a detector-emitter stage (sometimes called the 'video driver') is given in Fig. 10.6. The emitter-follower stage will require forward biasing and if d.c. coupling is used between the detector and emitter-follower the base bias must be prevented from upsetting the operation of the vision detector. It is thus usual to find d.c. feeds to both sides of the detector diode which ensure that 'anode' and 'cathode' of diode are at the same d.c. potential; or, alternatively, the 'anode' is slightly positive to the 'cathode'. In this case the diode is given a small forward bias to reduce distortion for small signal inputs—the technique used in Fig. 10.6.

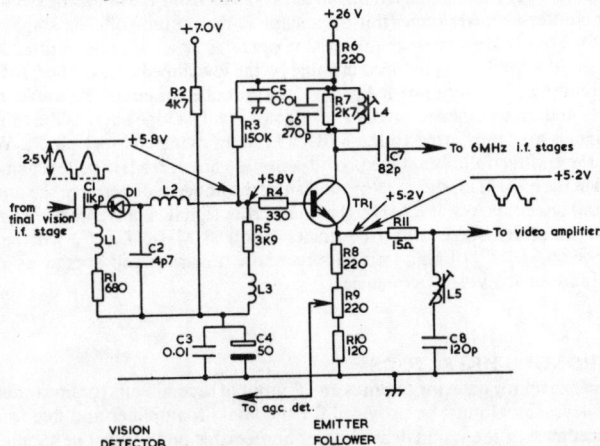

FIG. 10.6   USE OF DETECTOR EMITTER FOLLOWER

Forward bias for the emitter-follower TR_1 is supplied *via* R_2, L_3 and R_5 to TR_1 base. The d.c. at the lower end of R_2 is also fed to D_1 'cathode' *via* R_1 and L_1 whilst the 'anode' of the diode receives a d.c. potential *via* R_3. The resulting difference in potential across D_1 gives the diode a small forward bias. D_1 detects the negative half-cycles of the i.f. input which is capacitively coupled to the detector *via* C_1. After filtering of the detector output by the i.f. choke L_2, the video signal is applied to TR_1 base. Note that as the video signal goes towards sync. pulse level, the base bias of TR_1 is reduced. Since TR_1 acts as an emitter-follower for the video range of 0—5·5 MHz, the video signal at the emitter will follow that at the base but the amplitude will be slightly less (voltage gain of emitter-follower about 0·9). The output signal is developed across R_8, R_9 and R_{10} and a portion of the video signal is fed to the a.g.c. circuit from across R_9 which acts as the preset contrast control. The purpose of L_4, C_6 and L_5, C_8 is dealt with in the next section.

Intercarrier Sound Take-out

The intercarrier f.m. sound signal of 6 MHz resulting from the mixing of the vision and sound i.f.s in the vision detector must be extracted and fed to the 6 MHz i.f. stages prior to detection of the f.m. sound signal. The 6 MHz signal may be taken out directly from the vision detector or from a suitable point in the video stages of the receiver. Fig. 10.7 shows some common 6 MHz take-out arrangements used.

In diagram (a) the 6 MHz signal is extracted at the detector ouptut using a parallel tuned circuit L_1, C_1 which is resonant to 6 MHz. At this frequency the tuned circuit is of high impedance and most of the 6 MHz signal will be developed across it with very little appearing across the detector load R_1. At frequencies below 6 MHz the impedance of L_1, C_1 will be low and thus the tuned circuit does not interfere with the normal operation of the vision detector. Energy from L_1 is transferred to the 6 MHz i.f. amplifying channel *via* the coupling winding L_2.

The 6 MHz signal may be extracted from the detector emitter-follower stage and in diagram (b) it is taken out from the emitter circuit using a series tuned circuit L_1, C_1 which is resonant to 6 MHz. The intercarrier beat applied to the base of the transistor will also be present at its emitter. To 6 MHz, L_1, C_1 behaves as a low impedance, thus as far as 6 MHz signals are concerned the emitter is grounded and the intercarrier signal will not be passed on to the video amplifier stage. However, the reactance of L_1 will be high at the resonant frequency and the 6 MHz signal may be taken out from across L_1. At frequencies below 6 MHz, L_1, C_1 is of high impedance and the transistor operates as a normal emitter-follower stage to the video band of 0—5·5 MHz.

In diagram (c) the intercarrier signal is extracted from the collector circuit of the detector emitter-follower stage (this is the same as the emitter-follower stage shown in Fig. 10.6). To 6 MHz signals the transistor operates as a common-emitter amplifier with the emitter load being reduced in value by the low impedance of the 6 MHz series tuned circuit L_2, C_2 in series with R_4. L_1, C_1 in the collector circuit is parallel resonant to 6 MHz and thus amplified intercarrier signals are developed across the tuned load. These signals are transferred to the 6 MHz i.f. amplifying channel *via* C_3. With this method the emitter-follower stage provides useful gain at 6 MHz so less gain may be required in the 6 MHz i.f. stages. Note that since the impedance between the junction of R_4, R_6 and chassis is low at 6 MHz, the intercarrier signal is prevented from reaching the video amplifier stage. For frequencies below 6 MHz, L_1, C_1 will be of low impedance and L_2, C_2 of high impedance and the transistor will operate as a normal emitter-follower for video frequencies.

SYNCHRONOUS DETECTORS

A diode envelope detector requires an i.f. input of several volts for linear detection. The high level signal must be provided for the final i.f. amplifier and due to the low input impedance of the vision demodulator appreciable power must be supplied from the final i.f. amplifier. This leads to difficulty in the design of consistently stable i.f.

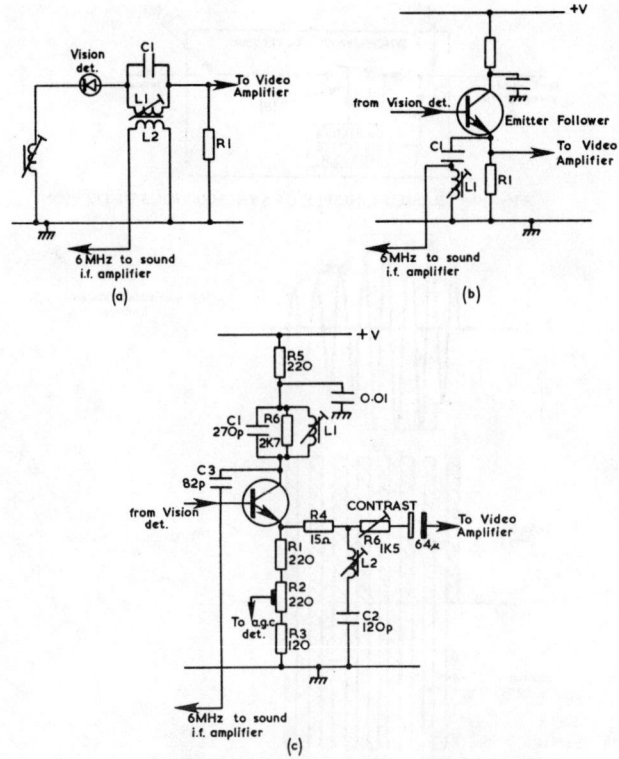

FIG. 10.7   6 MHz TAKE-OUT CIRCUITS

amplifiers when a diode demodulator is used because of the high level of signal present.

A synchronous or switched detector on the other hand is very linear in operation down to small input signals of the order of tens of millivolts. This allows the receiver gain to be shared between the i.f. stages and the video stages which lends itself to a more stable design.

Another advantage of the synchronous detector is that there is less mixing of the luminance, chrominance and sound components of the signal which produce unwanted frequency components detrimental to the performance of the receiver. In consequence less sound rejection is required in the i.f. amplifier and some reduction in the critical nature of fine tuning during colour reception is obtained.

Synchronous Demodulator Considered as a Switch

The synchronous demodulators used in receiver circuits function as switches and Fig. 10.8 shows the basic idea. Here the demodulator is considered as an ordinary electrical switch with the modulated vision i.f. carrier fed in at one side of the switch. The i.f. input is also fed to a limiting and selective amplifier which produces a square wave output at the vision i.f. of 39·5 MHz which is used to operate the switch.

It will be assumed that the switch is closed on positive half-cycles of the switching waveform *B* but open on the negative half-cycles as shown in Fig. 10.9. When the switch is closed the signal at the detector input is passed to the output. Thus the switch passes only the positive half-cycles of the modulated i.f. input to the output, *i.e.* the signal has been detected (waveform *C*). This demodulated output contains an

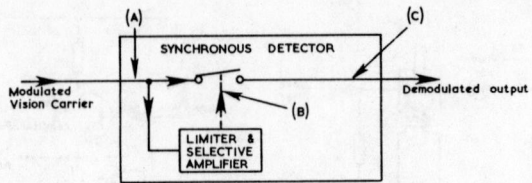

FIG. 10.8   BASIC PRINCIPLE OF SYNCHRONOUS DETECTION

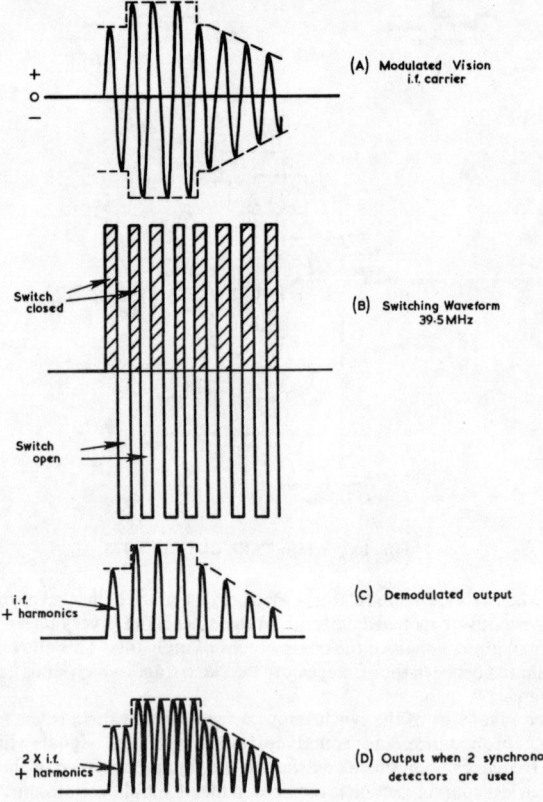

FIG. 10.9   WAVEFORMS EXPLAINING ACTION OF SYNCHRONOUS DETECTION

appreciable amount of the i.f. component since half-wave rectification has been used. The fundamental component of the i.f. may be reduced at the detector output by employing two such switching demodulators with their outputs combined. These are operated by opposite polarity switching waveforms and opposite polarity vision i.f. inputs, resulting in a full-wave output as shown in waveform D. There is now no output ripple at the fundamental i.f. frequency which assists in preventing instability. Thus a vision synchronous demodulator is a linear full-wave detector as opposed to the half-wave operation of the non-linear diode demodulator.

The vision synchronous detector should not be confused with the synchronous demodulators used in a colour receiver decoder which recover the colour-difference signals. These detectors require a separate oscillator for switching whereas in the vision synchronous detector the vision i.f. is used for the switching function. Synchronous detection of the vision signal has not been available in discrete component form because of the prohibitive cost but is now available in integrated circuit form.

Synchronous Demodulator I.C.

Fig. 10.10 shows the basic functions inside one type of synchronous demodulator i.c. The output of the final i.f. amplifier is fed to pin 7 of the i.c. with a peak-to-peak

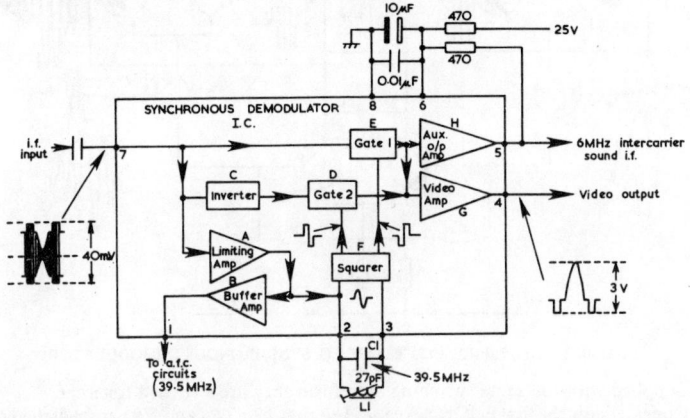

FIG. 10.10   INTERNAL FUNCTIONS OF I.C. SYNCHRONOUS DETECTOR

value of about 40 mV. A sample of the signal is amplitude limited in A and the output is used to develop a constant amplitude 39·5 MHz sine-wave across the tuned circuit L_1, C_1 (fitted externally to the i.c.)

This signal is squared in F and the anti-phase outputs are fed to the two gates D and E which constitute the synchronous detectors. Gate 1 is supplied with the i.f. input direct from pin 7 but gate 2 is fed with an antiphase i.f. input after inversion in C. The demodulated outputs of the two gates are combined, filtered and fed to video amplifiers H and G. The lower amplifier provides a video output of approximately 3 V peak-to-peak and the upper amplifier the 6 MHz intercarrier sound i.f. which requires filtering from the video signal. Pin 1 of the i.c. provides a constant amplitude 39·5 MHz vision i.f. carrier output *via* the buffer amplifier B. This output may be used to operate the a.f.c. circuits which correct drift and mistuning of u.h.f tuner local oscillator.

A simplified circuit diagram of the two gates is given in Fig. 10.11. Several emitter-followers and the biasing arrangements have been omitted. Also, in the i.c. R_3 and R_4 are replaced by constant current circuits but these simplifications do not alter the basic principle of operation.

TR_1 and TR_2 form a differential amplifier with their bases supplied with antiphase i.f. inputs. In practice it is necessary to feed only one base due to the common-emitter coupling *via* R_3, but it is easier to consider the operation with the bases fed in this way. TR_1 and TR_2 feed the input signals to the emitters of the switching transistors TR_3, TR_4, TR_5 and TR_6. The differential pair, TR_1 and TR_2, also feeds another differential pair TR_7 and TR_8 with the signal voltages from across R_1 and R_2. TR_7 and TR_8 produce antiphase signal voltages across the load resistors R_7 and R_8 and between these two points is connected the 39·5 MHz resonant circuit L_1, C_1 (connected externally to the i.c.). Diodes D_1 and D_2 are placed across the tuned circuit to act as limiters or squarers. Hence a constant vision i.f. voltage is produced across L_1, C_1 which is applied to the bases of TR_3, TR_4, TR_5 and TR_6 to switch them alternately ON and OFF. Common load resistors are used for these transistors; R_5 for TR_4 and TR_5 and R_6 for TR_3 and TR_6. It will be seen the switching waveform applied to TR_4 base is also fed to TR_6 base thus TR_4 and TR_6 switch ON together. Also, since TR_5 and TR_3

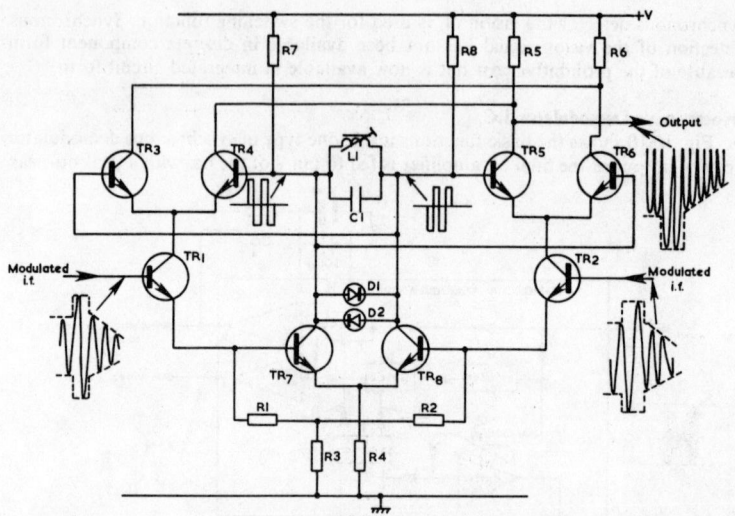

FIG. 10.11   SIMPLIFIED CIRCUIT OF I.C. SYNCHRONOUS DEMODULATOR

are supplied with the same switching waveform they are switched together.

Consider now the first half-cycle of the i.f. input when TR_4 and TR_6 are switched ON by the positive switching waveform at their bases. The i.f. input half-cycle being positive on TR_1 base causes TR_1 to conduct more and so an increasing current will flow in TR_4 and the load resistor R_5 with an amplitude proportional to the input. As regards TR_2 base input this is on a negative half-cycle and will cause the current in it to decrease. This decreasing current flows in TR_6 and the load resistor R_6. During the next half-cycle TR_3 and TR_5 are switched ON and the negative half-cycle of the i.f. input at TR_1 base causes a decrease in the current in TR_1 which flows in TR_3 and R_6. At the same time the positive half-cycle at TR_2 base causes an increase in TR_2 current which flows in TR_5 and R_5. Thus all half-cycles of the input which causes an INCREASE of current in TR_1 OR TR_2 flow in R_5 and those half-cycles which cause a DECREASE in current in TR_1 OR TR_2 flow in R_6. Note that for every full cycle of the i.f. input two half-cycles of current flow in the load resistors R_5 and R_6. If the voltage output from TR_5 collector is considered it will be seen that it is negative-going and resembles the output from a full-wave rectifier. After smoothing of the output the required video signal is obtained.

Comparison of Diode and Synchronous Demodulator Outputs

Consider the response of the two types of demodulators to the i.f. spectrum of frequencies given in Fig. 10.12.

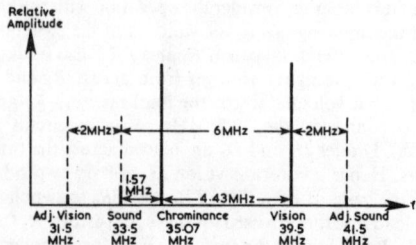

FIG. 10.12   AMPLITUDE-FREQUENCY DIAGRAM OF INTERMEDIATE FREQUENCIES

Ordinary Diode detector output

An envelope detector will produce the following outputs when fed with the various frequency components of Fig. 10.12.

(a)   The vision carrier modulation (the video signal): 0—5·5 MHz.

(b)   The adjacent channel vision carrier modulation.

(c)   Any amplitude modulation of the sound carrier or adjacent sound carrier.

In addition the detector output also contains the following beats:

(d)   Beat between vision and sound i.f. carrier: 6 MHz.

(e)   Beat between sound and chrominance i.f. carriers: 1·57 MHz.

(f)   Beat between vision and adjacent channel sound i.f.: 2·0 MHz.

Only ouputs (a) and (d) are required. The other components in the output represent interference which is detrimental to the performance of the receiver.

SYNCHRONOUS DETECTOR OUTPUT

A synchronous detector is a tuned detector operated by the switching frequency of 39·5 MHz. When fed with the components of Fig. 10.12 its outputs will be:

(g)   The vision carrier modulation (the video signal: 0—5·5 MHz.

(h)   Beat between vision and sound i.f. carriers: 6 MHz.

(i)   Beat between vision and adjacent sound i.f.: 2 MHz.

Of the above, (i) is the only component producing interference. There is no output due to (b), (c) and (e) as with the ordinary diode detector. Component (e), which produces an obtrusive pattern on vision, is most important and it means that less sound trapping is required in the i.f. stages (say −20 dB as opposed to −30 dB or more).

CHAPTER 11

VIDEO STAGES

T HE stages of the receiver that deal with either the detected luminance signal or the decoded colour information are all video stages, although in a colour receiver they are referred to as the luminance signal amplifiers and colour-difference signal amplifiers respectively.

The basic requirements of the video stages in a monochrome receiver, see Fig. 11.1, will be dealt with first.

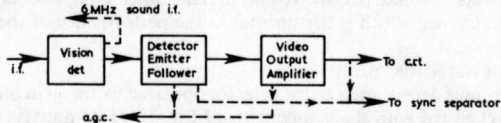

FIG. 11.1 VIDEO STAGES IN A MONOCHROME RECEIVER

As explained in Chapter 10, the output of the vision detector is usually fed to the video output stage *via* an emitter follower to prevent the low input impedance of the common emitter video output stage loading the detector, thereby ensuring maximum linearity of its output signal. The emitter-follower stage does not provide any voltage gain, therefore the output amplifier must develop sufficient video signal voltage to drive the c.r.t. The low output impedance of the emitter-follower stage is, however, suitable for driving the output amplifier. A low drive impedance is necessary as the input capacitance of the common-emitter output amplifier is rather high. This capacitance arises mainly out of the internal feedback within the transistor (Miller effect). In addition to driving the c.r.t. the video stages must also supply feeds to the sync. separator stage and a.g.c. circuit. These feeds may be taken from either the emitter follower or the output stage.

Video stages handling the luminance signal must satisfy the following requirements.

Frequency Response

To preserve the fine picture detail conveyed by the luminance signal the video amplifier must operate over a wide frequency range. Ideally, the gain should be uniform over the frequency range of 0 Hz—5·5 MHz, as shown in Fig. 11.2. To maintain the gain down to 0 Hz a d.c. amplifier is required, *i.e.* d.c. coupling must be used throughout from the vision detector right up to the modulating electrode of the c.r.t. Sometimes a.c. coupling is used or partial d.c. coupling, in which case the d.c. component is removed or only partly retained.

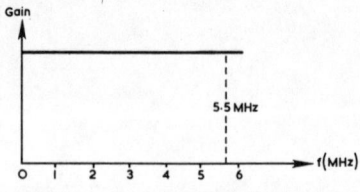

FIG. 11.2 IDEAL FREQUENCY RESPONSE

Phase Response

The relative phases of all frequency components present in the video signal must be preserved to avoid phase distortion of the video content. If an amplifier does not introduce phase shift this requirement is satisfied. Practical amplifiers, however, exhibit some phase shift particularly at high and low frequencies. To avoid phase

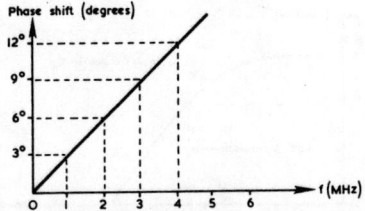

FIG. 11.3 PHASE SHIFT PROPORTIONAL TO FREQUENCY

distortion any phase shift introduced should be proportional to frequency as illustrated in Fig. 11.3, since this is the condition for constant delay at all frequencies. For example, a 3° phase shift of a 1 MHz sine-wave represents a timing error of approximately 8·34 ns relative to a 1 MHz sine-wave with zero phase shift. To produce the same timing error of 8·34 ns for a 2 MHz sine-wave, a phase shift of 6° is required.

Gain

A peak-to-peak video drive of about 60 V on average is required to drive a monochrome c.r.t. (depending upon the d.c. operating conditions) from black to peak white. The vision detector normally gives out a peak-to-peak video signal of between 1—2 V, thus the video amplification required is of the order of 30—60. This voltage gain may be obtained from a single transistor stage.

BASIC VIDEO OUTPUT STAGE

A basic circuit of a common emitter transistor video output stage commonly used is shown in Fig. 11.4. Essentially, a video amplifier is a resistance loaded amplifier where the load R_L is used to couple the amplified video signal to the c.r.t. Only small power is

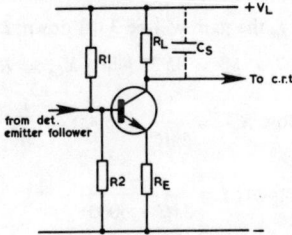

FIG. 11.4 BASIC TRANSISTOR VIDEO AMPLIFIER

needed to drive the c.r.t. thus the video output stage acts primarily as a voltage rather than a power amplifier. The voltage gain of 30—60 must be maintained up to 5·5 MHz, so an r.f. type transistor is used. Commonly, video output transistors have a cut-off frequency above 100 MHz and the intrinsic bandwidth of the transistor does not impose a limit on the upper video frequency gain of the stage. An undecoupled emitter resistor (R_E) is used which provides current negative feedback to stabilise the parameters of the transistor. Base bias is provided by the potential divider R_1, R_2 but the d.c. bias may be obtained directly from the emitter-follower stage.

The voltage gain (A_v) of the amplifier at low frequencies is approximately given by

$$A_v = \frac{R_L}{R_E}$$

when h_{fe} is high. Therefore, with constant values for the collector and emitter resistors it would be expected that the gain would remain uniform over the video frequency range. Unfortunately, however, the gain falls at high video frequencies due to stray capacitance across R_L (see Fig.11.5). This stray capacitance (C_s) is made up of the

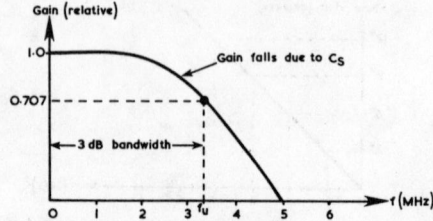

FIG. 11.5   RESPONSE OF NON-COMPENSATED VIDEO AMPLIFIER

capacitance of the c.r.t., wiring and output capacitance of the transistor and the input capacitance of the sync. separator stage (if fed from the collector of the output stage).

At high video frequencies, the falling reactance of C_s shunts R_L thereby reducing the effective load and causing the gain to fall. Because C_s is in parallel with R_L, when the reactance of C_s falls to a value such that it equals the value of R_L the gain of the stage will fall to

$$0.707 \frac{R_L}{R_E}$$

i.e. the gain will be 3 dB down. To give some indication of the frequency at which this will occur take the following representative values: $R_L = 5$ kΩ, $R_E = 100$ Ω and $C_S = 10$ pF. The gain at low frequencies will be

$$\frac{R_L}{R_E} = \frac{5000}{100} = 50.$$

At a particular frequency f_u the gain will be 3 dB down, *i.e.*

$$0.707 \times 50 = 35.35 \text{ when } X_{Cs} = R_L.$$

$$\text{Now } X_{C_s} = \frac{1}{2\pi f C} \text{ or } 5000 = \frac{1}{2\pi f C}$$

$$\text{therefore } f = \frac{1}{2\pi C \times 5000}$$

$$\text{or } f = \frac{10^{12}}{6.284 \times 10 \times 5000}$$

$$\text{thus } f \simeq 3.2 \text{ MHz.}$$

Thus in this case the 3 dB bandwidth would be 0—3·2 MHz which is insufficient to meet the requirement of 0—5·5 MHz. One way of increasing the 3dB bandwidth is to reduce the value of R_L. If a collector load of 2·5 kΩ is used the 3 dB bandwidth would be doubled to 6·4 MHz but the gain would be halved to 25 (see Fig. 11.6) since the reactance of C_s must fall to 2·5 kΩ to cause a 3 dB gain reduction. The reduced gain of the stage may still be sufficient, but the smaller value of collector load may cause the maximum power rating of the output transistor to be exceeded which will now be considered.

Assume that the transistor is operating in class–A (this is used in an a.c. coupled stage) and that a peak-to-peak video output of 120 V is required. The video output voltage requirement sets the minimum value of the d.c. supply that may be used. For the video output mentioned a d.c. supply of, say, 160 V would be used. These

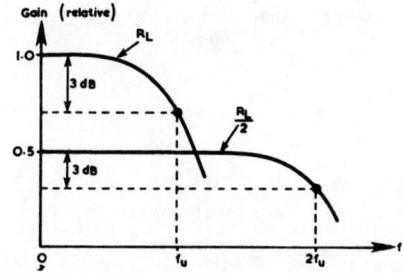

FIG. 11.6   EFFECT OF LOAD RESISTOR VALUE ON FREQUENCY RESPONSE

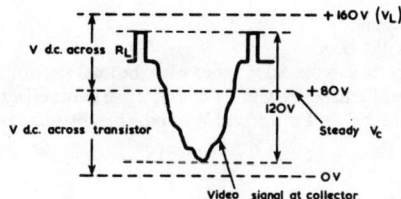

FIG. 11.7   CLASS-A OPERATION

conditions are shown in Fig. 11.7.

Now the MINIMUM VALUE of R_L may be found from

$$R_L = \frac{(V_L)^2}{4P_{max}},$$

where V_L is the d.c. supply and P_{max} is the maximum power rating for the transistor.

With a P_{max} of 1 W (typically) and a V_L of 160 V

$$R_L = \frac{(160)^2}{4} = 6.4 \text{ k}\Omega.$$

For a 3 dB bandwidth of 0—5.5 MHz and with a C_s of 10 pF, the MAXIMUM VALUE of R_L that can be used is

$$X_{C_s} = \frac{1}{2\pi fC} = \frac{10^{12}}{6.284 \times 5.5 \times 10^6 \times 10} \simeq 2.91 \text{ k}\Omega$$

It is clear, therefore, that an R_L value of 2.9 kΩ which satisfies the frequency requirement cannot be used without exceeding the P_{max} rating of the transistor. To avoid exceeding P_{max} the value of R_L must be higher than 6.4 kΩ. Suppose a value of 7 kΩ is chosen to allow for circuit tolerances. The collector current of the transistor is given by

$$\frac{V_L}{2R_L} = \frac{160}{2 \times 7000} \simeq 11.5 \text{ mA}$$

If the peak-to-peak input voltage is, say, 2 V, the voltage gain required is $\dfrac{120}{2} = 60$.

Since $A_v = \dfrac{R_L}{R_E}$, the value of R_F will be $\dfrac{7000}{60} \simeq 117\Omega$. The upper video frequency f_u will be

$$f_u = \frac{1}{2\pi C X_{C_s}}$$

$$= \frac{10^{12}}{6.284 \times 10 \times 7000}$$

$$\simeq 2.28 \text{ MHz}.$$

Various h.f. compensating techniques may be used to improve the response at high frequencies.

Inductance Compensation

(i)   SHUNT PEAKING COIL

Here a small inductor L is placed in series with the load resistor R_L as shown in Fig. 11.8(a). The presence of L in series with R_L causes a rise in the effective collector circuit load impedance at the h.f. end of the video band, thereby increasing the gain of the

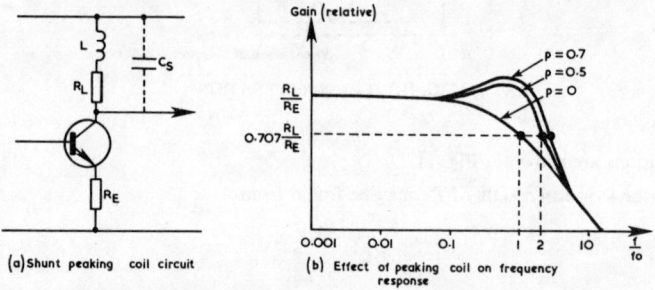

(a) Shunt peaking coil circuit

(b) Effect of peaking coil on frequency response

FIG. 11.8   INDUCTANCE COMPENSATION (SHUNT PEAKING COIL)

amplifier and partly compensating for the falling reactance of C_s. To establish a suitable value for L the following design formula is often used

$$L = p \, R_L^2 \, C_s \text{ H (where } p \text{ is a number usually between } 0.3 \text{ and } 0.7)$$

The effect on the response of the amplifier of varying the factor p is shown in Fig. 11.8(b) using 'generalised' curves that can be applied to any amplifier. On the logarithmic horizontal axis the ratio f/f_o is used, where f_o is the frequency at which the gain has fallen to 0.707 of the low frequency gain in an uncompensated amplifier and f is the frequency under consideration. When $p = 0$, L is zero and we have the curve for the uncompensated amplifier. At 3 dB down on this curve $f/f_o = 1$. If a value of $p = 0.5$ is taken it will be seen that at 3 dB down, $f/f_o = 2$, i.e. the -3 dB video frequency will be twice that of an uncompensated amplifier. With an upper -3 dB frequency of 2.28 MHz (using the example of the previous section) and a p value of 0.5, the -3 dB upper video frequency will be increased to 4.56 MHz which is a useful improvement. Taking the previous values of $R = 7$ kΩ and $C_s = 10$ pF with a p value of 0.5, the approximate value of L will be

$$L = p R_L^2 C_s$$

$$= 0.5 \times (7000)^2 \times 10 \times 10^{-12} \text{H}$$

$$= 245 \ \mu\text{H}.$$

Values for L are commonly of the order of 50—300 μH. Once the approximate value of L has been established using the design formula, various values of L around the calculated value are experimented with using pulse test signals until the desired video response is achieved.

(ii) SERIES PEAKING COIL

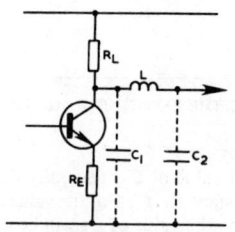

FIG. 11.9 SERIES PEAKING COIL CIRCUIT

Here a small inductance L is placed in series with the output signal path as opposed to being in shunt with it as in the shunt peaking coil circuit. If the inductor is fitted close to the collector connection, it splits the stray capacitance into two parts: C_1 the output capacitance of the transistor, and C_2 the input capacitance of the c.r.t. plus wiring capacitance. C_1, L and C_2 now take the form of a low-pass filter. The value of L is chosen so that the filter passes all frequency components up to the desired upper video frequency with minimum attenuation but provides increasing attenuation above the upper limit of the video band.

As with the shunt peaking circuit, 'generalised' curves may be used to establish the required amount of compensation needed but the design is generally more complex. For best performance C_1 and C_2 must be in a definite ratio and in practice the final value of L is arrived at by using pulse testing methods.

Compensation Using n.f.b.

Compensation for the fall in gain at high video frequencies due to the effects of C_s may be achieved using frequency selective n.f.b. In the uncompensated circuit, R_E is undecoupled and introduces n.f.b. (d.c. and a.c.) but the degree of feedback is the same for all frequencies within the video band. The voltage gain is then settled by the ratio of the collector to emitter load resistors. If a capacitor (C_E) of suitable value is placed across R_E as in Fig. 11.10 it is possible to arrange that the amount of n.f.b. at h.f. is small compared with the degree of n.f.b. at the l.f. and m.f. ranges of the video band.

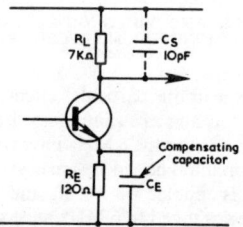

FIG. 11.10 COMPENSATING USING FREQUENCY DEPENDENT NEGATIVE FEEDBACK

The value of C_E must be such that at l.f. and m.f. its reactance is high compared with R_E value so that the effective emitter load for signals is R_E. Whereas at the h.f. end of the video band (due to the falling reactance of C_E) the effective emitter impedance is reduced thereby causing a reduction in the amount of n.f.b. and a rise in voltage gain. The idea is shown in Fig. 11.11 where at the h.f end of the video band, the reduced feedback just compensates for the falling gain due to C_s thereby giving a level response

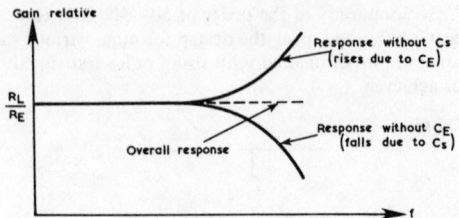

FIG. 11.11   SHOWING THE EFFECT OF C_E ON FREQUENCY RESPONSE

up to a higher video frequency.

As a guide to the required value of C_E, generally the time-constant of R_E, C_E is made the same as the time-constant R_L, C_s. For the values given in Fig. 11.10, the value of C_E is about 580 pF. Usually, the value of C_E will lie in the range of 300—5000 pF.

Typical Circuit Arrangements

One example of a video amplifying channel is given in Fig. 11.12. This uses a driver stage TR_1 and a video output amplifier TR_2 with a.c. coupling between stages. The maximum peak-to-peak video at TR_2 collector is 110 V and this requires about 2 V of video signal from the detector.

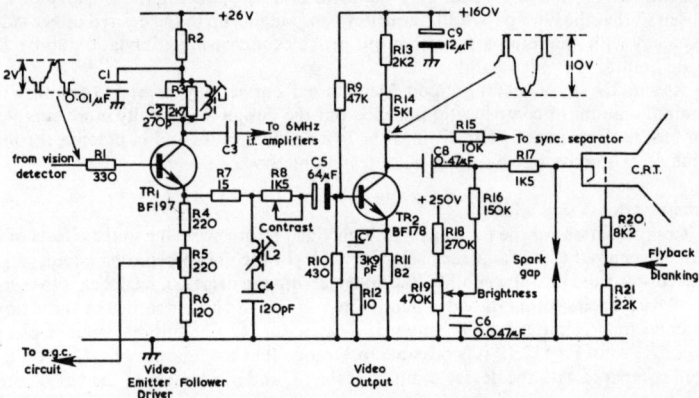

FIG. 11.12   EMITTER-FOLLOWER AND VIDEO OUTPUT STAGE USING A.C. COUPLING (MONOCHROME RECEIVER)

The operation of TR_1 stage in this particular circuit was described in Chapter 10 using Figs. 10.6 and 10.7(c). TR_1 acts as a common collector stage to video signals but as a common emitter arrangement to the 6 MHz intercarrier signal which is extracted by L_1, C_2 in the collector circuit. The video signal at TR_1 emitter, which is less in amplitude than at the base, is coupled *via* R_7, R_8 and C_5 to the base of the output transistor. L_2, C_4 which is series tuned to 6 MHz rejects the intercarrier sound signal from the output stage thereby preventing a fine dot pattern being displayed. R_8 serves as the manual contrast control by forming one arm of an attenuator with the input impedance of TR_2; as the resistance of R_8 is reduced the contrast is increased. C_5 couples the video signal to TR_2 base but blocks its d.c. component. Although theoretically the d.c. component should be retained, its omission is a matter of opinion as in practice the results on the picture are not very detrimental.

TR_2 is biased to class–A by the potential divider R_9, R_{10}. The collector load is formed by R_{13}, R_{14} and R_{11} is the emitter resistor. High frequency compensation is provided by C_7 in the emitter circuit. R_{12}, which is part of the compensation network, sets a limit on the minimum emitter impedance and hence the maximum gain at high video frequencies.

Class–A operation of the output stage is shown in Fig. 11.13. For the circuit values given, a standing collector current slightly in excess of 10 mA is indicated which

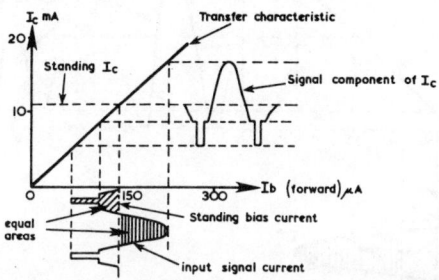

FIG. 11.13 CLASS-A OPERATION OF VIDEO OUTPUT AMPLIFIER

requires a standing base current of about 150 μA for the BF178 transistor. Note that due to the a.c. coupling (via C_5) the input current will be balanced about the standing base bias current as shown. Thus on the excursions of the input towards the sync. pulse tips the collector current will reduce, whereas when the input goes towards the peak of the video content the collector current will increase. A peak-to peak collector current of about 15 mA will be needed to develop 110 V of signal at the collector. Note also that as the collector current increases, the collector voltage reduces, thus the output voltage is negative-going on the video content as is required for cathode modulation of the c.r.t.

Consider now the video output supplied to the c.r.t. This is fed to the tube cathode via C_8 and R_{17}. C_8 blocks the d.c. component and R_{17} helps to prevent voltage spikes entering TR_2 stage (where they may cause damage) in the event of an e.h.t. flashover within the c.r.t. In this receiver brightness control is introduced by varying the steady voltage at the c.r.t. cathode, the grid being connected to chassis potential via R_{20} and R_{21}. The brightness voltage is supplied from R_{19} which varies the d.c. at the slider connection over the range 0—+160 V approximately. As the brightness voltage is made more positive, the c.r.t. beam current is reduced, i.e. the brightness of the picture is decreased. R_{16} stands off the brightness control from the video signal thereby preventing R_{19} (or C_6) shorting out the video signal. At the minimum setting of R_{19}, the c.r.t. cathode voltage will not be zero due to the beam current (high) which flows in R_{16}. C_6 decouples R_{19} to signals.

The a.c. coupled video signal 'sits' about the brightness potential from R_{19} and the correct setting for the brightness control is shown in Fig. 11.14 (upper waveform). The brightness control is adjusted so that blanking level on the video drive corresponds to zero beam current, i.e. black is represented by zero beam current. If the brightness control is increased or decreased from this ideal setting, the video drive waveform and its mean level (the brightness potential) will move bodily to the right or left on the diagram. If moved to the left, zero beam current will correspond to a grey on video drive waveform and the peak white level will result in a decrease in beam current, i.e. the picture will be too dark overall. Conversely, if moved to the right, black level on the video drive will correspond to grey in the picture, i.e. the picture will be too light overall. Although the brightness setting shown in the diagram is technically the correct one, the contrast range on the screen is a matter of individual preference and is affected by the ambient lighting in the room.

Altering the contrast setting will cause either a larger or smaller amplitude video signal to be applied to the c.r.t. If larger (increased contrast) the beam current will increase on the peaks of the video signal; if smaller (reduced contrast) the beam current peaks will be less. With a.c. coupling, however, alteration of the contrast setting or picture

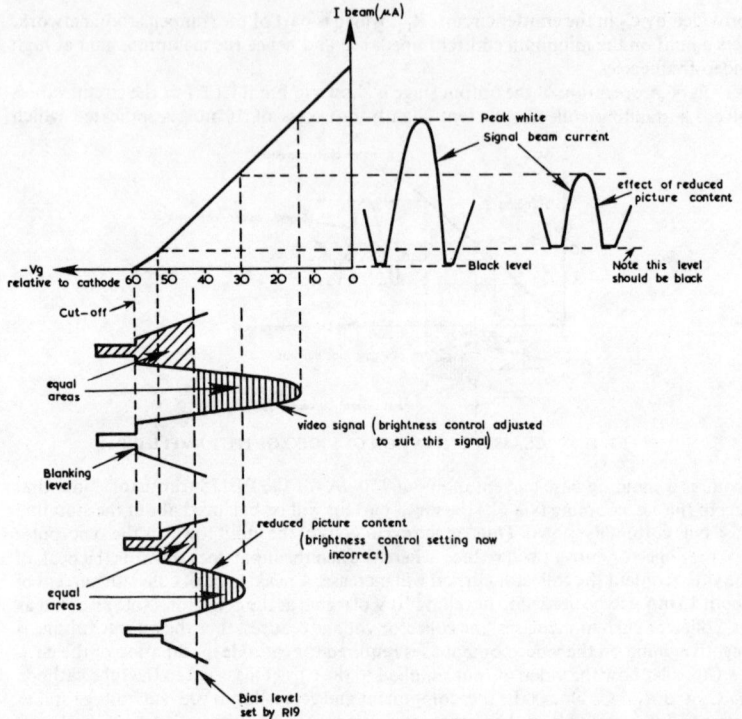

FIG. 11.14 SETTING OF BRIGHTNESS CONTROL POTENTIAL FOR A.C. COUPLED VIDEO
SIGNAL

content (see lower waveform) will require an adjustment of the brightness setting to
ensure that blanking level on the drive waveform corresponds to zero beam current.

Another example of a video amplifying channel arrangement is given in Fig. 11.15
but here d.c. coupling is used throughout. Detected video and 6 MHz intercarrier sound
from the vision detector D_1 are developed across the load R_4 and fed to the base of the
emitter-follower stage TR_1. C_3, L_1 in the emitter circuit of TR_1 form a series tuned
circuit resonant to 6 MHz. Due to the low impedance of this circuit at resonance, very
little intercarrier signal is passed on to the TR_2 stage. However, because of the
relatively high reactance of L_1 at 6 MHz, appreciable 6 MHz signal is developed across
this inductor which thus provides the take-out point for the intercarrier signal. Video
signal from across R_5 is coupled to TR_2 base *via* the contrast control R_7. As the slider of
R_7 is moved to the right a smaller video signal is applied to TR_2 and the picture contrast
is lowered. R_8, R_9 are used to minimise the change in d.c. at TR_2 base with variations in
the setting of R_7, thereby ensuring a constant black level. TR_2 is the video output stage
providing a peak-to-peak output of about 70 V. R_{12} is the collector load and R_{10} the
emitter resistor. H.F. compensation is provided by the shunt peaking coil L_2 in the
collector circuit and by C_4 in the emitter circuit. The negative-going video signal is
applied to the c.r.t. cathode *via* the parallel rejector L_3, C_5. This circuit is tuned to 6
MHz and rejects any residual intercarrier signal.

As d.c. coupling is used in the video stages we must now consider the effect of the
d.c. component of the detected video signal, Fig. 11.16. In diagram (a) we have the
modulated i.f. waveform which is normally shown with a zero reference level. If the
negative half-cycles are detected and the detector load resistor is connected on one side
to chassis potential (as is often shown in basic detector circuits), the video output will

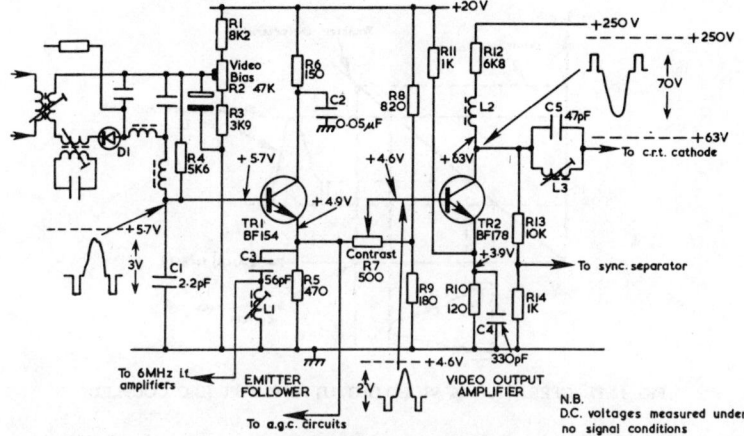

FIG. 11.15 EMITTER-FOLLOWER AND VIDEO OUTPUT AMPLIFIER USING D.C. COUPLING (MONOCHROME RECEIVER)

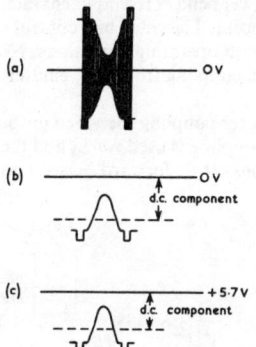

FIG. 11.16 EFFECT OF D.C. COMPONENT

be as in diagram (b). Note here that the d.c. component or average value of the signal is negative with respect to zero. When the video output is d.c. coupled to the emitter-follower stage, the video is no longer referenced to zero but to the base bias potential of the emitter-follower stage. Thus, in diagram (c) the zero reference of diagram (b) is now the base bias potential of the emitter-follower stage which in Fig. 11.15 is $+5.7$ V. The d.c. component is, therefore, negative with respect to the base bias potential. If the positive half-cycles of the input had been detected, the d.c. component would have been positive with respect to the standing base potential.

Returning now to the circuit of Fig. 11.15 the standing bias on TR_1 stage must be set high so that the input signal excursions towards the sync. pulse tips reduce the bias. Note that the d.c. measurements, if taken under signal conditions, will be less than the 'no signal' values given on the diagram as the meter will indicate the average value of the signal waveforms.

Since the signal at TR_1 emitter 'follows' the signal at its base and the emitter output is directly coupled to TR_2 base, the d.c. component of the signal at TR_2 base will be negative with respect to the standing base potential of TR_2. Thus the forward bias for TR_2 stage is initially set high as shown in Fig. 11.17. The high standing bias current will therefore give rise to a high standing collector current of about 20 mA in this case (compare with Fig. 11.13). As the signal input progresses towards the sync. pulse tips

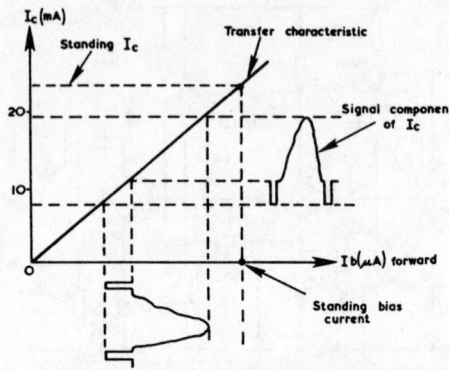

FIG. 11.17   OPERATION OF VIDEO OUTPUT AMPLIFIER (D.C. COUPLED)

the current in TR_2 is reduced. The designer has to ensure linear operation for the video information without limiting the signal on peak white. Also, as the input signal takes the operation towards the lower bend of the input characteristic, excessive 'crushing' of the sync. pulses should not occur. The video bias control (R_2) is used to set the forward bias of TR_2 stage for optimum operating conditions. Note that R_2 adjusts TR_1 base potential and due to the d.c. coupling from TR_1 emitter to TR_2 base it also sets TR_2 base bias.

The circuit arrangement for coupling the video output of TR_2 stage to the c.r.t. is shown in Fig. 11.18. D.C. coupling is used via L_3 and the beam current limiting diode D_2. Under normal conditions, D_2 is forward biased and the video signal at the c.r.t.

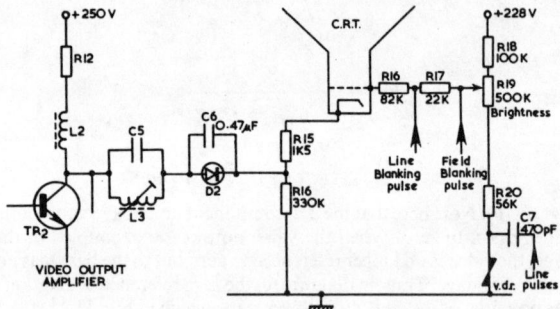

FIG. 11.18   D.C. COUPLING BETWEEN VIDEO OUTPUT AND C.R.T.

cathode is clamped to the output at TR_2 collector. During high drive conditions when the c.r.t. beam current exceeds a predetermined value, D_2 becomes reverse biased due to the rise at its cathode as a result of the beam current flowing in R_{16}. The video signal is then a.c. coupled to the cathode via C_6 thereby limiting the beam current to a safe level.

A different arrangement is used for brightness variation in this circuit. Here the brightness voltage is applied to the c.r.t. grid instead of the cathode (as with the previous circuit). The brightness voltage from R_{19} is set so that the grid potential is negative with respect to the positive potential applied to the cathode from the collector of TR_2. Increasing the voltage setting of R_{19} reduces the grid-to-cathode bias, thereby increasing the beam current and so increasing the picture brightness. Negative-going pulses of suitable duration and amplitude are fed to the grid to black out the line and

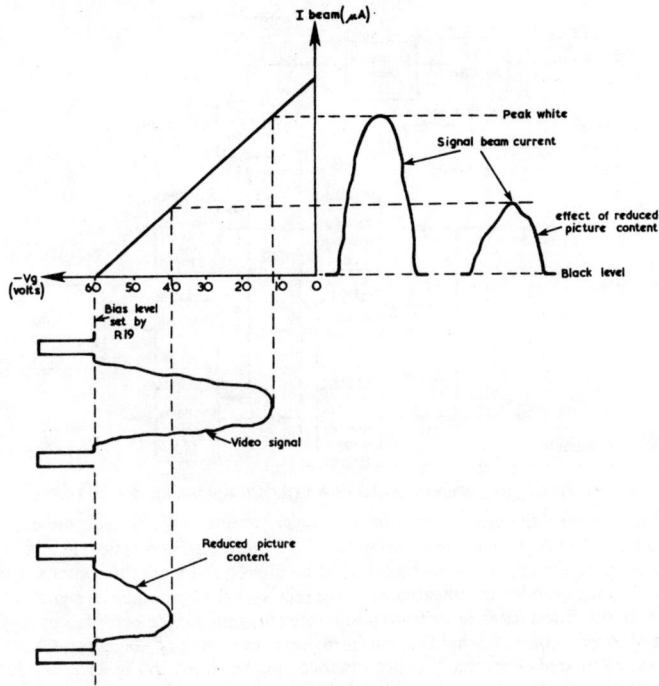

FIG. 11.19   CORRECT SETTING OF BRIGHTNESS CONTROL POTENTIAL FOR D.C. COUPLED
VIDEO SIGNAL

field flyback.

The correct setting of the brightness control potential is shown in Fig. 11.19. Here the brightness potential is set so that the video signal excursions are always on one side of this voltage level and the sync. pulses take the operation of the c.r.t. beyond beam cut off. With d.c. coupling when the picture content alters the black level of the picture remains constant (compare with Fig. 11.14).

Video Stages (Colour Receiver)

One possible arrangement of the video stages when using primary signal drive to the three guns of the colour display tube is given in Fig. 11.20. Primary signal (RGB) drive is the most common way of driving a colour tube in a solid-state receiver and this method will be used to provide a basic understanding of the video signal paths.

All the stages represented by the blocks are VIDEO stages but are usually given the designations shown. Any video stage that handles the luminance signal information should have a video frequency requirement of 0—5·5 MHz, *i.e.* blocks (1), (2), (3), (4), (8), (9) and (10). Stages such as blocks (5), (6) and (7) which deal with the colour-difference signals, require a response of 0—1 MHz only, since the colour-difference signals are restricted in bandwidth compared with the luminance signal. The video stages in a colour receiver use the same basic circuit techniques as a monochrome receiver. Of course, the high frequency restrictions placed on the wide band luminance stages do not apply to colour-difference amplifiers and often no h.f. compensation is necessary in these stages.

Luminance Signal Path

The luminance signal amplifying channel is formed by blocks (1), (2), (3), (4) and the matrixing amplifiers (8), (9) and (10). As with the monochrome receiver, outputs are

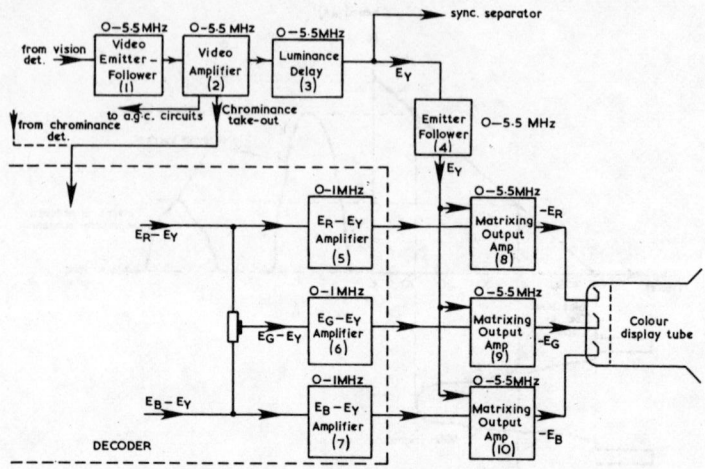

FIG. 11.20   THE VIDEO STAGES IN A COLOUR RECEIVER (R.G.B. DRIVE)

necessary to feed the sync. separator and a.g.c. circuits and although these may be taken from the points indicated, variations will be met with in practice. In addition to these outputs, the chrominance signal must be filtered out from the composite signal and fed to the decoder as indicated. In some receivers the chrominance signal is taken out from the vision detector or from a separate chrominance detector fed by way of a stage of chrominance i.f. amplification from one of the vision i.f. stages (see Chapter 9).

One method of extracting the chrominance signal is shown in Fig. 11.21(a). TR_1 is a video stage in the luminance channel with R_1 serving as the collector load and R_3 as the

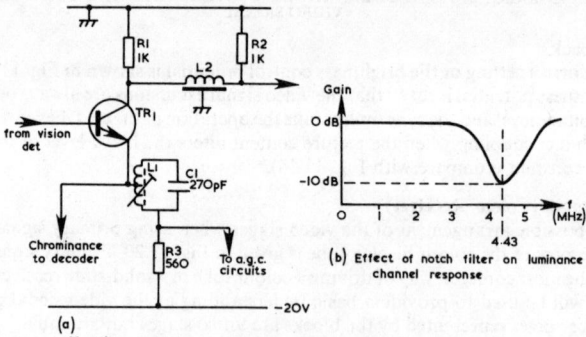

FIG. 11.21   LUMINANCE CHANNEL VIDEO AMPLIFIER WITH NOTCH FILTER AND LUMIN-
ANCE DELAY LINE

emitter resistor. No special h.f. compensation is used in this stage as the gain requirements are modest so a small value of collector load resistance may be employed. L_1, C_1 in the emitter circuit is parallel tuned to 4·43 MHz and acts as a notch filter. At 4·43 MHz its impedance will be high causing a large degree of negative feedback. In consequence, the gain of TR_1 is reduced at the chrominance signal frequencies which therefore do not appear to any large extent across R_1. Due to the high impedance of the filter at resonance it provides a suitable take-out point for the chrominance signals which are then fed to the decoder section of the receiver. The effect of the notch filter on the response of the luminance channel is shown in diagram (b). It will be seen that luminance signals around 4·43 MHz are removed from the luminance channel which degrades the high frequency response to some extent but is necessary to prevent the chrominance signals modulating the raster *via* the luminance channel. In some colour receivers the notch filter is removed automatically when receiving monochrome signals. The automatic facility may be achieved by fitting a diode across the filter which is reversed biased by a voltage from the colour killer stage in the decoder allowing the filter to operate normally on colour transmissions. During monochrome transmissions when the colour killer voltage disappears, the diode becomes forward biased thereby shorting out the notch filter and improving the response of the luminance channel. The Q of L_1, C_1 is maintained by the tap on L_1 and the bandwidth of the filter is ± 1 MHz centred on 4·43 MHz.

Because the luminance and colour-difference signals pass through video stages with different bandwidths without any special compensation they will arrive at the c.r.t. at different times. Any signal passing through a circuit with restricted bandwidth is delayed. Owing to the narrower bandwidth of the colour-difference signal amplifiers, the colour-difference signal is delayed more than the luminance signal. The practical effect of this is that the colour information would be displayed slightly to the right of the luminance information on the screen (about 0·5 cm on a 22″ screen). This tends to cause transitions in the picture to be blurred. To prevent misregistration of the colour and luminance contents of the picture, the luminance signal is given a small delay of about 0·6 μs.

One way of delaying a signal is to send it along a transmission line. The amount of delay for a given length depends upon the distributed inductance and capacitance of the line. Most forms of transmission lines, *e.g.* coaxial cable, have a small inductance and capacitance per unit length. To reduce the length of the cable needed for a specific delay, special forms of construction may be used which increase the distributed inductance and capacitance. The form of construction used for the luminance delay line is given in Fig. 11.22. Here, a single layer of wire is wound over a copper foil strip which is laid on the outer surface of an insulating tube. The inductance of the wire and

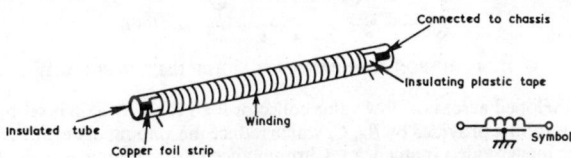

FIG. 11.22   CONSTRUCTION OF LUMINANCE DELAY LINE

its relatively high capacitance to the metal foil provide the necessary delay while at the same time keeping the overall dimensions to reasonable proportions (about 9 cm in length by 1 cm diameter). The line must be terminated by resistors at either end to matched the characteristic impedance of the line which usually lies in the region of 1·2 kΩ (see R_1 and R_2 of Fig. 11.21).

A different arrangement for extracting the chrominance signal is given in Fig. 11.23. The composite signal from the vision detector emitter-follower stage is fed through R_1 and C_1 to a bandpass filter comprising L_1, C_2—C_3—C_4—L_2, C_5. The response of the filter is shown in Fig. 11.24. L_1, C_2 is parallel resonant to 6 MHz and rejects the intercarrier sound signal from the luminance stage TR_1 and the chrominance take-out. L_1, C_2 also resonates as a series tuned acceptor circuit with C_3 providing a lift in gain at 5·4 MHz. L_2, C_5 form a parallel resonant circuit giving a rise in gain at 3·4 MHz. Also, with C_4 they form a series acceptor to attenuate luminance signals around 2·5 MHz from the chrominance take-out. The overall response is double peaked with almost a flat top symmetrical about 4·43 MHz and high rejection at 2·5 MHz and 6 MHz. Thus the filtered signal available at the junction of C_3, C_4 contains the chrominance sidebands and colour burst which are fed to the decoder.

So far as the input to TR_1 base is concerned this comprises the full luminance bandwidth except the d.c. component (blocked by C_1) and the intercarrier signal (rejected by L_1, C_2). The luminance signal is amplified and inverted by TR_1 and the

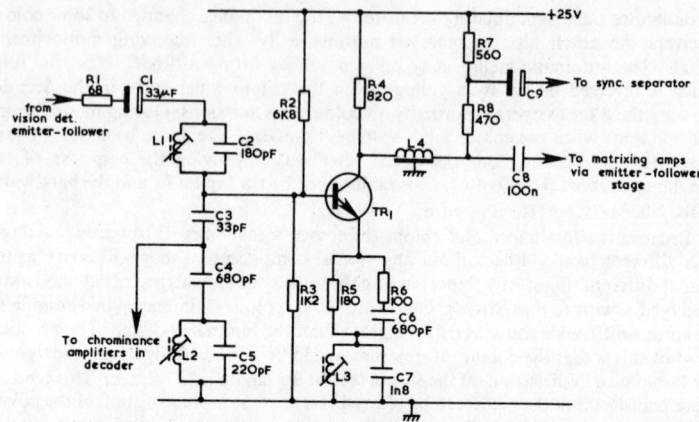

FIG. 11.23   LUMINANCE CHANNEL VIDEO AMPLIFIER WITH BANDPASS FILTER AND
LUMINANCE DELAY LINE

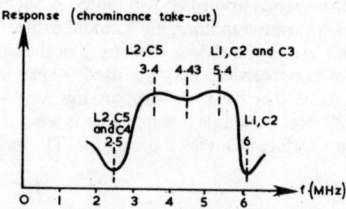

FIG 11.24   RESPONSE OF BANDPASS FILTER USED IN FIG. 11.23

output is developed across the low value collector load R_4. TR_1 gain is set by R_5 and
h.f. compensation is provided by R_6, C_6 which reduce the amount of negative feedback
towards the higher video frequencies. Chrominance information is rejected from the
luminance channel by the parallel tuned circuit L_3, C_7 in the emitter circuit. This circuit
is tuned to 4·43 MHz and, due to its high impedance at resonance, it introduces a large
amount of negative feedback thereby reducing the gain of TR_1 to chrominance signals.
L_4 is the luminance delay line which equalises the mean delay of the luminance channel
with that of the chrominance channel. The input of the delay line is terminated by R_4
and the output by R_7, R_8. A suitable level of signal from across R_7 is fed to the sync.
separator *via* C_9. Delayed luminance signal is fed to the matrixing amplifiers *via* C_8 and
an emitter-follower stage.

Colour-Difference Amplifiers and Output Matrix

An example of an amplifying channel for one of the colour-difference signals ($E_B -$
E_Y) is given in Fig. 11.25. As the circuits for the ($E_R - E_Y$) and ($E_G - E_Y$) channels are
similar they need not be considered.

The output of the U synchronous demodulator in the decoder is fed to TR_1 base *via*
C_1 and R_2. TR_1 is a common emitter video amplifier with R_4 serving as the collector
load. As the upper video frequency is only 1 MHz no special h.f. compensation is
necessary. R_3 sets the gain of the stage and a suitable portion of the $E_B - E_Y$ signal is
fed to the $E_G - E_Y$ amplifier *via* R_5 and C_2 for matrixing with an appropriate amount
of the $E_R - E_Y$ signal. The amplified and inverted blue-difference signal from TR_1
collector is fed to an emitter-follower TR_2. This stage drives TR_4 on its emitter with a
$-(E_B - E_Y)$ signal *via* the blue gain preset R_9. TR_4 and TR_5 constitute the output
matrixing amplifier. The transistors are connected in series to share the high d.c. supply

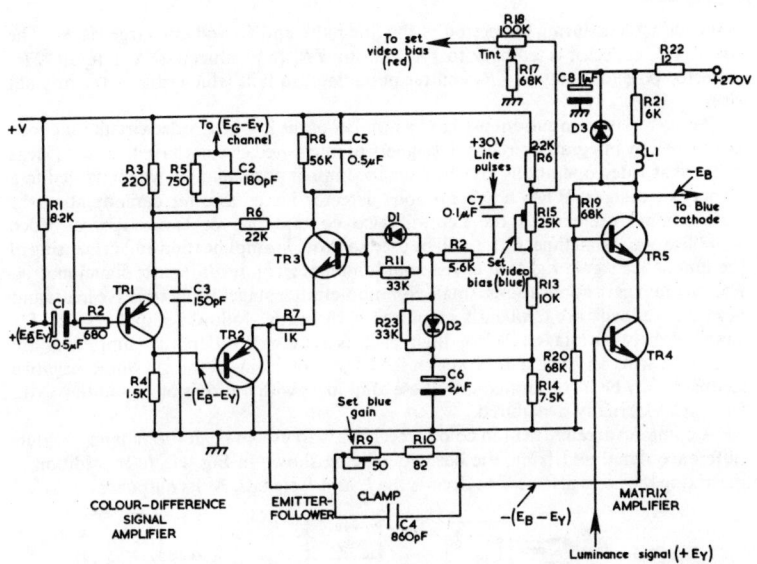

FIG. 11.25 COLOUR-DIFFERENCE AMPLIFIER AND OUTPUT MATRIXING AMPLIFIER

which is necessary to provide a signal of 150 V peak-to-peak at the c.r.t. cathode. R_{21} is the load resistor and L_1 (shunt peaking coil) provides h.f. compensation to maintain the video frequencies up to 5·5 MHz. D_3 prevents the collector of TR_5 from rising above the supply line potential in the event of ringing in L_1. Base bias for TR_5 is from the potential divider R_{19} and R_{20} and TR_4 receives its base bias from the source of the luminance signal input.

A $+E_Y$ signal is fed to TR_4 base and a $-(E_B - E_Y)$ signal is fed to the emitter. Since a $-(E_B - E_Y)$ signal at the emitter is equivalent to a $+(E_B - E_Y)$ at the base, the effective base drive is $+E_Y + (E_B - E_Y) = +E_B$. Now a $+E_B$ signal on TR_4 base will produce a $-E_B$ signal at TR_5 collector which is of the right sign (polarity) to drive the c.r.t. on its cathode. On monochrome, when the colour difference signals disappear, $+E_Y$ signal at TR_4 base will give rise to a $-E_Y$ signal at TR_5 collector with the result that the c.r.t. is driven by the luminance signal only as is required.

The clamp transistor TR_3 is used to eliminate d.c. drift in the video stages and the a.c. couplings (C_1 for example) which otherwise might cause tint changes in the picture. Note that TR_4 emitter is d.c. coupled via R_{10} and R_9 to TR_2 and this stage is d.c. coupled to TR_1. Any change in TR_2 emitter potential as a result of d.c. variations in TR_1 or TR_2 stages will alter TR_4 emitter potential, hence the steady potential at TR_5 collector which will affect the bias on the blue gun. The luminance signal fed to TR_4 is d.c. restored to a level set by the brightness control and this is stabilised so preventing drift at TR_4 base.

TR_2 and TR_4 emitters are clamped to a preset level during the line blanking period by a line pulse applied via C_7 to D_1 and D_2. This 30 V positive pulse causes D_2 and D_1 to conduct, thereby connecting TR_3 base to the positive potential present at the junction of R_{13}, R_{14}. When TR_3 comes ON the current flowing is determined by the voltage across C_6 (set by the video bias control) and the d.c. potential at TR_2 emitter (decided by d.c. drift). Suppose that TR_2 emitter has drifted in the negative direction. This will cause TR_3 to conduct more and its collector potential to fall to a lower voltage (as C_5 charges). Due to the d.c. coupling from TR_3 collector to TR_1 base, TR_1 will conduct harder causing its collector voltage to rise (with respect to chassis). TR_2 base and hence TR_2 emitter follow this rise which tends to counteract the original fall, thus stabilising TR_2 and TR_4 emitter potentials. Should TR_2 emitter drift in the positive direction, TR_3

will conduct less during the period of the line pulse and C_5 will discharge *via* R_8. The rise at TR_3 collector is fed back to TR_1 causing TR_1 to conduct less. As a result, TR_1 collector potential falls so TR_2 emitter potential also falls which cancels the original rise.

Many receivers in current use have a number of the low level video circuit functions performed by integrated circuits. Designers of integrated circuits have their own ideas as to what video operations may be packaged into a particular i.c. which has led to a variety of designs. When a synchronous detector i.c. is used for demodulating the vision signal quite often the i.c. will also contain one or two stages of video amplification. Sometimes an i.c. will be used to provide amplification and processing of the luminance signal right up to the final luminance/colour-difference signal matrix. For medium gain applications, single common-emitter stages using resistive loads and negative feedback are frequently employed with emitter-follower amplifiers used for coupling between stages. When high gain is required, differential amplifiers are employed with an emitter-follower to feed the signal out of the i.c. Some negative feedback may be incorporated into these amplifiers with a provision for adding extra feedback externally if required.

A common arrangement in colour receivers is to use an i.c. for luminance/colour-difference signal matrixing, the basic idea being shown in Fig. 11.26. In addition to matrixing the i.c. is also used to decode the U and V signals. At its outputs on pins 2, 3

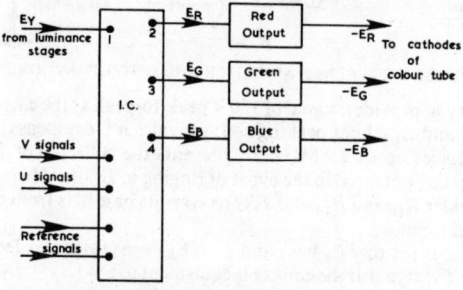

FIG. 11.26 USE OF I.C. FOR LUMINANCE COLOUR-DIFFERENCE SIGNAL MATRIXING (SIMPLIFIED DIAGRAM)

and 4 primary signals are available which are amplified up to a suitable level to drive the c.r.t. using discrete transistor output amplifiers. The essential processes of the i.c. in carrying out the matrixing of the luminance and colour-difference signals is shown in Fig. 11.27. Here the three colour-difference signals and the luminance signal are fed to a resitive matrix which provides the required primary signals at its output. These are then fed out of the i.c. *via* emitter-follower stages. The diagram also shows a resistive matrix employed for developing the green difference signal from the decoded red and blue difference-signals.

One of the primary signal output stages of Fig. 11.26 is given in Fig. 11.28. As the circuit is the same for the E_R, E_B and E_G outputs, only the E_R amplifier has been shown. The $+E_R$ output from pin 2 of the i.c. is developed across R_1 which is the emitter load for the emitter-follower in the i.c. The signal is d.c. coupled to the base of the video output transistor with R_2, C_1 forming a filter to remove unwanted harmonics. Amplified and inverted signal is developed across the collector load R_3 and fed to the 'red' cathode *via* the series peaking coil L_1 (damped by R_{11}) and R_{12} (flashover protection). In addition to the h.f. compensation provided by L_1 extra boost for the high frequencies is given by R_6, C_2 (using negative feedback) to maintain the gain of the stage up to 5·5 MHz. R_4 divides down the h.t. supply to the stage so that TR_1 'sees' an effective voltage of 190 V.

The emitter circuit of TR_1 is in the form of a bridge circuit, operating in an almost balanced condition. R_8, R_9 and R_{10} form one side of the bridge with TR_1 and R_5

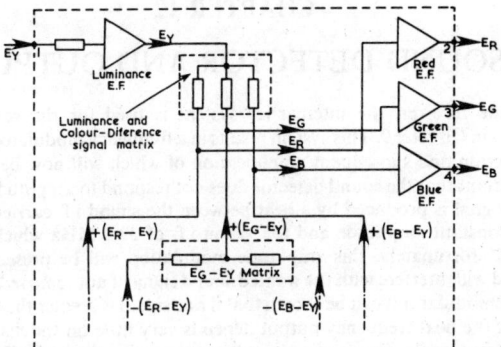

FIG. 11.27   SCHEMATIC OF PART I.C. SHOWING VIDEO SIGNAL PATHS AND MATRIXING
NETWORK

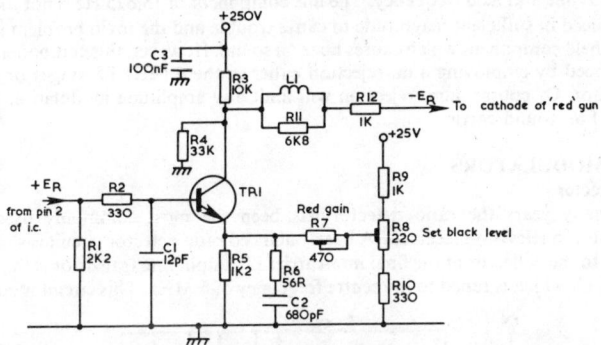

FIG. 11.28   RED OUTPUT (REPEATED FOR BLUE AND GREEN OUTPUTS)

forming the other. R_7 (the 'red' gain control) is connected between the two mid points
which are balanced. Thus very little current flows in R_7 and in consequence adjustment
of the control has little effect on the d.c. conditions of TR_1. When the video signal
swings towards peak white (on monochrome with only the luminance signal applied)
the emitter voltage of TR_1 rises. This causes the bridge to go out of balance and R_7
varies the load 'seen' by the emitter and hence the gain of the stage. By this
arrangement, R_7 may be adjusted to give the correct drive to the c.r.t. (when setting up
the grey-scale) with very little effect on the d.c. conditions and so little change in
background colour. R_8 adjusts the balance of the bridge to accommodate component
tolerances and sets the collector voltage of TR_1 to the required d.c. conditions at black
level.

CHAPTER 12

SOUND DETECTOR AND OUTPUT

I N a 625-line receiver, the intercarrier system is used for the sound signal as discussed in Chapter 9. This system results in a frequency modulated i.f. of 6 MHz, the detection and subsequent amplification of which will now be described.

It is important that the sound detector does not respond to amplitude modulation. The 6 MHz signal is produced by a beat between the sound i.f. carrier of 33·5 MHz which is of constant amplitude and the vision i.f. of 39·5 MHz which is amplitude modulated. Unfortunately, this amplitude modulation will be present on the beat frequency and will interfere with the wanted sound signal if not removed. If two signals are fed to a demodulator it can be shown that if one signal is greater than the other, the magnitude of the beat frequency output depends very little on the magnitude of the larger of the two signals. This is one reason for attenuating the 33·5 MHz sound i.f. so that at all times the vision i.f. of 39·5 MHz is much greater than the sound i.f. carrier. The main components of the amplitude modulation transferred from the vision i.f. carrier lie at line and field frequency. The line component of 15,625 Hz is not likely to be reproduced in sufficient magnitude to cause trouble and the main problem is with the 50 Hz field component which causes buzz on sound. However, this component may be suppressed by employing a.m. rejection either in the 6 MHz i.f. stages or in the demodulator. Of course, a.m. rejection will limit any amplitude modulation of the incoming f.m. sound carrier.

F.M. DEMODULATORS
Ratio Detector

For many years the ratio detector has been the most commonly used f.m. demodulator in television receivers. A basic balanced ratio detector circuit is shown in Fig. 12.1. In the collector of the final intercarrier i.f. amplifying transistor is the tuned circuit L_1, C_1 which is tuned to the centre frequency of 6 MHz. This circuit is coupled

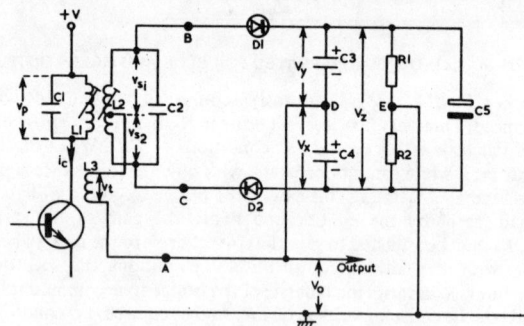

FIG. 12.1 BASIC RATIO DETECTOR CIRCUIT

by mutual inductance to the secondary circuit L_2, C_2 which is also tuned to 6 MHz. L_3 is a tertiary winding which is mutually inductively coupled to the primary inductor L_1.

The circuit works on the principle that at the centre frequency the voltage induced into the secondary is 90° out of phase with the primary voltage (or the voltage in the tertiary winding). As the frequency deviates according to the modulation the voltages are no longer at 90° but are either greater or less than 90°. Consider the phasor diagram of Fig. 12.2(a) at the centre frequency. Since L_1, C_1 is at resonance to the signal collector current of the transistor, the voltage across the primary v_p will be in phase with the collector current i_c. The current that flows in L_1 (i_{L1}) will lag v_p by 90° (assuming a pure inductance). The current in L_1 will induce a voltage v_{L2} into the

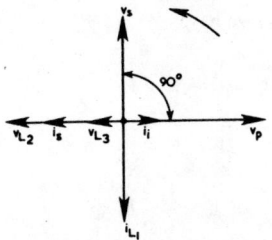

(a) Phasors at centre frequency — voltage induced in secondary v_s leads the primary voltage v_p by 90°

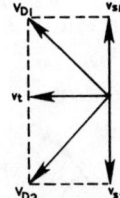

(b) Resultant voltage V_{D1} and V_{D2} applied to the circuit at the centre frequency

FIG. 12.2   PHASORS AT THE CENTRE FREQUENCY

secondary winding L_2, also a voltage v_{L3} into the tertiary winding L_3. These voltages will lag the current i_{L1} by 90° i.e. they will act in opposition to the applied e.m.f. v_p. The voltage v_{L2} induced into L_2 will drive a current i_s around the secondary circuit L_2, C_2 which acts as a series tuned circuit as regards this e.m.f. Since the circuit is at resonance, i_s will be in phase with v_{L2}. The secondary current flow through C_2, hence the voltage v_s across C_2 and hence across L_2 will lag the current by 90°. Thus v_s and v_p are in phase quadrature as shown.

Now v_s can be divided into two voltages v_{s1} and v_{s2} which, as regards the centre-tap, are 180° out of phase [see Fig. 12.2(b)]. The voltage induced into the tertiary winding v_{L3} will be designated v_t and this provides the reference phase from the primary. This e.m.f. is in phase quadrature with v_{s1} and v_{s2} as shown. The effect of these voltages on the circuit operation will now be considered. Between points B and C on the circuit diagram there are two diodes, D_1 and D_2 in series with equal value capacitors C_3 and C_4 and equal value resistors R_1 and R_2. The effective voltage between points A and B (V_{D1}) is the vector sum of v_{s1} and v_t, whereas the effective voltage between points A and C is the vector sum of $|v_{s2}|$ and v_t. At the centre frequency V_{D1} and V_{D2} are of the same magnitude. The voltage V_{D_1} is applied to D_1 which will cause the diode to conduct once per cycle charging C_3 with the polarity shown. The voltage V_{D_2} is applied to D_2 which conducts each half-cycle causing C_4 to charge with the polarity as shown. As V_{D1} and V_{D2} are of the same magnitude and $C_3 = C_4$, the voltages across the capacitors will be of the same magnitude. Let the voltage across $C_4 = V_x$ and the voltage across $C_3 = V_y$. Also, let the total voltage across R_1 and $R_2 = V_z$. The output voltage from the circuit (V_0) is taken between points D and E.

Since $R_1 = R_2$

$$V_o = \frac{V_z}{2} - V_y \qquad (12.1)$$

$$\text{or } V_o = V_x - \frac{V_z}{2} \qquad (12.2)$$

Adding equations (**12.1**) and (**12.2**) we have

$$2V_o = \frac{V_z}{2} - V_y + V_x - \frac{V_z}{Z}$$

$$2V_o = V_x - V_y$$

$$\text{or } V_o = \frac{V_x - V_y}{2}.$$

As C_3 and C_4 are of the same value, at the centre frequency $V_y = V_x$ and V_o is zero.

If the frequency of the input to L_1, C_1 changes, *i.e.* when the carrier is modulated, v_s and v_l depart from the phase quadrature condition. This is shown in Fig. 12.3(a) where it is assumed that the deviation causes the carrier input to increase in frequency. L_1, C_1

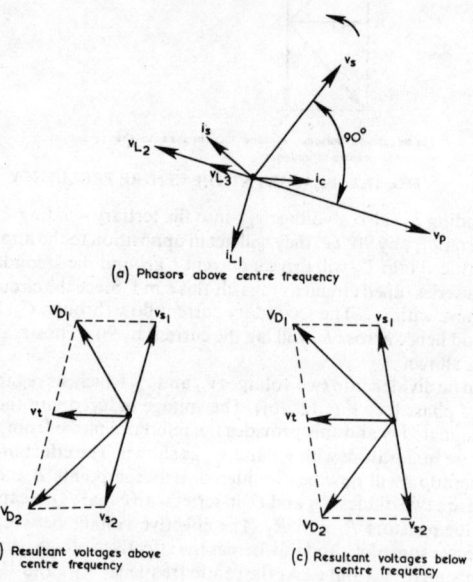

(a) Phasors above centre frequency

(b) Resultant voltages above centre frequency

(c) Resultant voltages below centre frequency

FIG. 12.3 PHASORS WITH DEVIATION APPLIED

is then no longer resonant and a larger current flows in C_1 than in L_1 thus the circuit behaves as a capacitor. Therefore, the voltage across the primary v_p will lag i_c. The current i_{L_1} in L_1 will lag v_p by 90° and this current will cause e.m.f.s v_{L_2} and v_{L_3} to be induced in L_2 and L_3 both of which will lag i_{L_2} by 90°. The e.m.f. v_L will drive a current i_s around the secondary circuit but since a series circuit appears inductive above resonance, the current will lag v_{L2}. The voltage across C_2 (and L_2) will lag i_s by 90°. As a result, v_s leads v_p by an angle less than 90°. If deviation is considered in the opposite direction it will be found that v_s leads v_p by an angle greater than 90°. As regards the effective voltages applied to the circuit for operation above and below the centre frequency these will now appear as in Fig. 12.3(b) and (c).

Above the centre frequency $V_{D2} > V_{D1}$, thus D_2 will conduct harder than D_1, and C_4 will receive a greater charge than C_3, *i.e.* V_X will be greater than V_y. However, the total voltage V_z remains approximately constant (assuming that the amplitude of the input does not vary).

Suppose that $V_2 = 3$ V and $V_y = 2$ V, therefore

$$V_o = \frac{V_x - V_y}{2}$$

$$= \frac{3 - 2}{2} = +\tfrac{1}{2} \text{ V}.$$

There is now a net voltage between D and E (D positive with respect to E), hence there will be an output from the circuit.

Below the centre frequency $V_{D1} > V_{D2}$, thus D_1 conducts harder than D_2 and C_3 receives a greater charge than C_4, i.e. V_y will be greater than V_x. Suppose that $V_y = 3$ V and $V_x = 2$ V, therefore

$$V_o = \frac{V_x - V_y}{2}$$

$$= \frac{2 - 3}{2} = -\tfrac{1}{2} \text{ V}.$$

Again there is an output voltage but this time D is negative with respect to E. Thus as the carrier deviates from its centre frequency an output is obtained from the circuit with an amplitude proportional to the amount of deviation and with a polarity dependent on the direction of the deviation. With sinusoidal modulation of the carrier, a sine-wave at the modulation frequency will appear between D and E. It will be seen that the ratio of the voltages V_y/V_x varies with the modulation hence the reason for the name 'ratio detector'.

Amplitude Limiting

An important feature of the ratio detector circuit is that it will provide amplitude limiting of the input signal. This is the purpose of the large value capacitor C_5 (4—25 μF). The capacitor will become charged to the mean value of the voltage across R_1 and R_2 and any rapid variations due to amplitude modulation of the carrier from noise or the vision signal will have little effect on this voltage. If the carrier amplitude suddenly increases, C_5 attempts to charge to a higher level *via* the diodes and any resistance placed in series with the diodes (see Fig. 12.4). For a fall in carrier amplitude C_5 will discharge *via* R_1 and R_2. In both cases, due to the large value of capacitor used, the charge and discharge time-constants are relatively long, say 0·1 s (discharge) and 0·01 s (charge). Thus the output V_0 does not respond to amplitude variations of the input, only frequency variations due to the modulation. The circuit will operate with any reasonable input, and the voltage across C_5 adjusts to the input level. As a receiver is brought closer to the correct tuning point, the voltage across C_5 increases. The voltage across C_5 thus serves as a useful indication of best tuning when carrying out i.f. alignment.

Practical Arrangement

A practical balanced ratio detector circuit is shown in Fig. 12.4. The effectiveness of a.m. rejection depends upon the balance of the circuit which is the reason for the inclusion of R_1 and R_2 in series with the diode circuits. These resistors assist in balancing the characteristics of the two diodes to prevent unequal diode currents (at the centre frequency) for large signal inputs. R_1 is adjusted for minimum a.m. in the output. C_6 serves as the a.f. load capacitor and completes the circuit to chassis as

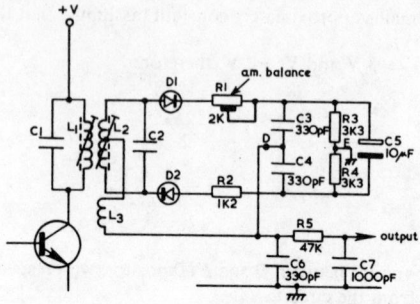

FIG. 12.4   PRACTICAL BALANCED RATIO DETECTOR CIRCUIT

regards the i.f. R_5, C_7 form a de-emphasis network to correct for pre-emphasis given to the signal at the transmitter. This network should provide a time-constant of 50 μs (47 μs in this case). Instead of connecting point E to chassis, point D may be so connected and the output taken from point E. The operation is identical.

A typical relationship between output voltage and frequency is given in Fig. 12.5. If there is to be zero distortion the output voltage should be proportional to frequency

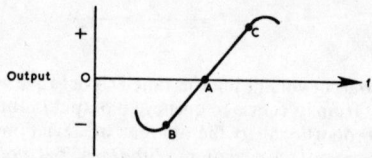

FIG. 12.5   CHARACTERISTIC OF RATIO DETECTOR CIRCUIT

deviation and this is seen to be true over the range of B to C. These two points, of course, should be far enough apart on the frequency scale to cover the maximum deviation. Since the deviation we are concerned with is ± 50 kHz, points B and C must be at least 100 kHz apart, preferably, say, 150 kHz. Point A corresponds to the unmodulated carrier of 6 MHz.

A number of unbalanced circuits are possible and one is given in Fig. 12.6. In this case there will be the normal audio frequency output but also half the steady voltage

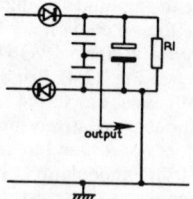

FIG. 12.6   UNBALANCED RATIO DETECTOR CIRCUIT

across R_1 at the centre frequency. Provided that the d.c. output is not fed to the a.f. amplifier, it has no effect on the operation of the demodulator or a.f. amplifier.

Quadrature Coincidence Detector

Integrated circuits are now being used for the sound demodulator but an i.c. does not lend itself to the ratio detector circuit and a new principle is used. It is hardly worth while just using the i.c. for the demodulator, so the i.c. performs other functions, e.g. i.f. amplifier, volume control and a.f. amplifier, as shown in Fig. 12.7 for the TBA750Q i.c.

The i.f. amplifying section of the i.c. was discussed in Chapter 9. It is, however, worth noting at this point that the high gain of the i.f. amplifier causes a limiting of the

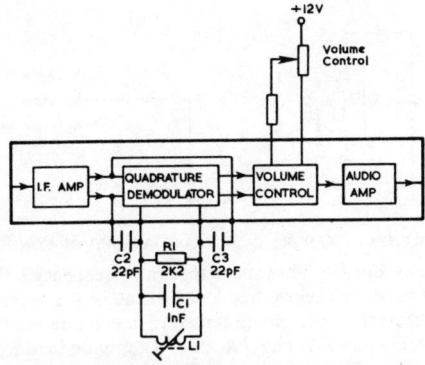

FIG. 12.7   MAIN FUNCTIONS OF TBA 750Q SOUND I.C.

signal input to the demodulator section. Thus the input to the demodulator is a clipped sine-wave signal at the i.f.

Turning now to the demodulator section of the i.c., a simplified circuit is shown in Fig. 12.8, known as a 'quadrature coincidence detector'. Transistors TR_1 and TR_2 form a long-tail pair with R_2 as the common-emitter resistor, which in practice is

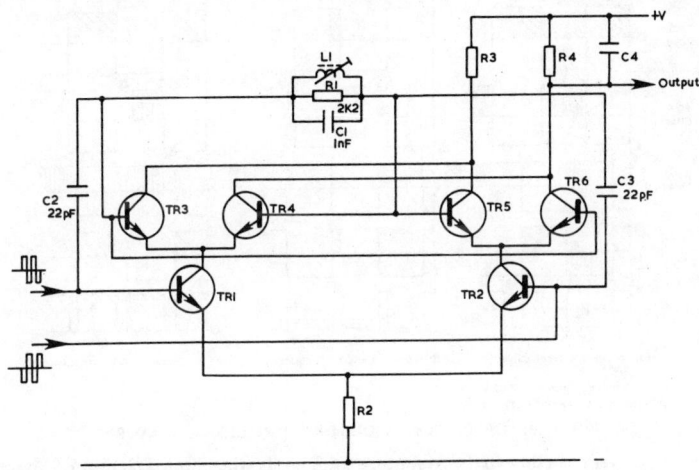

FIG. 12.8   INTEGRATED CIRCUIT QUADRATURE COINCIDENCE DEMODULATOR

usually replaced by a constant-current circuit. The bases of TR_1 and TR_2 are fed with opposite polarity signals from the limiting i.f. amplifier so that the inputs are of square-wave form. Opposite polarity inputs are not essential, only one base need be fed. A tuned circuit L_1, C_1 is connected across the antiphase inputs through small capacitors C_2 and C_3. These components are connected externally to the i.c. although in some i.c.s C_2 and C_3 are formed by reverse biased diodes on the integrated circuit. L_1, C_1 are tuned to 6 MHz thus at the centre frequency the tuned circuit behaves as a resistance. The voltage across L_1, C_1 (V_0) is in phase with the current i (see Fig. 12.9). However, since the reactance of C_2 and C_3 at 6 MHz is large compared with the resistance of the tuned circuit, the current i leads the input voltage V_i by almost 90°. Thus at the centre frequency V_i and V_o are in phase quadrature. Above the centre frequency when the tuned circuit acts capacitively i leads V_o, but below the centre frequency i lags V_o as the

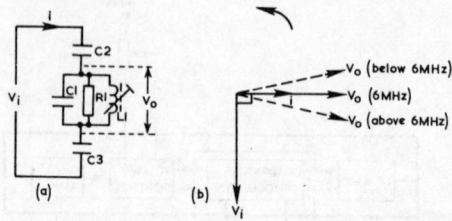

FIG. 12.9   PHASORS SHOWING QUADRATURE CONDITION

tuned circuit is then inductive. Thus above the centre frequency V_o leads V_i by an angle less than 90° and below the centre frequency the angle is greater than 90°.

The voltage across the tuned circuit is used to switch transistors TR_3 and TR_6 also TR_4 and TR_5. Transistors TR_3 and TR_5 have a common load R_3 and TR_4 and TR_6 share the load R_4. The effects of TR_3 and TR_5 on the current in the load R_3 may be neglected as this output is not used. The antiphase voltages from either side of the tuned circuit are sufficient in amplitude to switch TR_4 and TR_6 from fully ON to the fully OFF. At the centre frequency these voltages are 90° out of phase with the voltages applied to TR_1 and TR_2 bases. The operation may be explained using the waveforms of Fig. 12.10(a) where perfect square waves have been shown for simplicity. As the inputs to TR_1 and TR_2 are also of sufficient amplitude to switch TR_1 and TR_2 from fully ON to

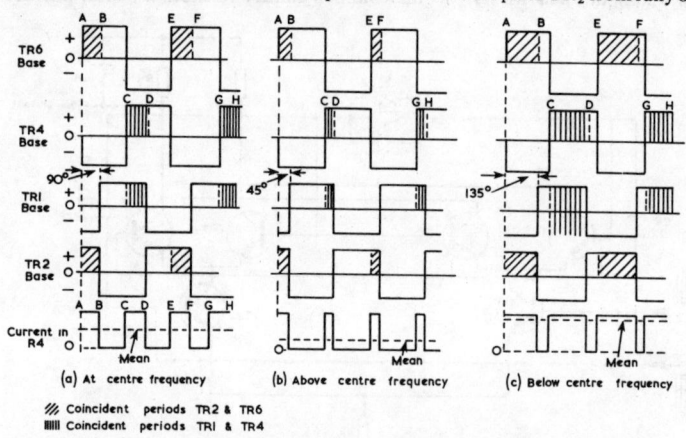

FIG. 12.10   OPERATION OF QUADRATURE DEMODULATOR

fully OFF, current can flow in the common load R_4 only when both TR_1 and TR_4 bases are positive or when TR_2 and TR_6 bases are positive. The coincident periods are shown in the diagram, A–B and E–F for TR_2/TR_6; C–D and G–H for TR_1/TR_4. As a result the current in R_4 flows for the periods shown, where the mean value is half the peak value (equal mark and space periods).

Suppose now that the carrier is deviated and that the voltage across the tuned circuit is 45° out of phase with the input as in diagram (b). The coincident periods for TR_2/TR_6 and TR_1/TR_4 are now considerably reduced. In consequence, the current flowing in R_4 lasts for much shorter periods and the mean value of the current is less than when undeviated. If deviation is considered in the opposite direction as in diagram (c), where the phase difference is 135°, the coincident periods are increased. Current then flows in R_4 for longer periods and the mean value of this current is greater than when undeviated.

It will be seen that the mean value of the voltage across R_4 will vary with the phase and that this phase changes with the deviation. Therefore, after smoothing out the pulses across R_4, the voltage will vary with the deviation as is required. The relationship between deviation and output can be made very linear over a range which is determined by the damping of the tuned circuit (R_1). The advantages of this type of demodulator are.

(a) There is no fundamental component of the i.f. in the output which helps to maintain stability. The ripple in the output which is at twice the i.f. may be removed by a capacitor (C_4) connected external to the i.c. across R_4. This capacitor, if of the correct value, produces the required de-emphasis.

(b) Only a single tuned-circuit is required therefore alignment is made easy.

(c) Since the demodulator is a switching circuit the output depends only on the frequency deviation and not on the magnitude of the input (provided it is large enough to switch properly) and so the circuit automatically suppresses a.m.

The actual integrated circuit is more complex than that shown. Constant current circuits are used in place of the emitter resistor and emitter-followers are employed to reduce the loading between stages. The principle of operation is, however, unaffected by these additions.

Audio Signal Stages

The de-emphasised signal output from the f.m. demodulator is now fed to the audio stages of the receiver. As this section of a t.v. receiver serves the same function as the corresponding section of a radio receiver similar circuits are used. Class-A output stages are normally used to reduce h.t. current variations and 'sound on vision' effect. Both transformer and transformerless output stages may be used and often integrated circuits are employed.

One example of the audio stages found in a modern t.v. receiver are shown in Fig. 12.11. An i.c. is used to provide the volume control and preamplifier functions with a discrete transistor used for the output stage. Obviously, an external volume control

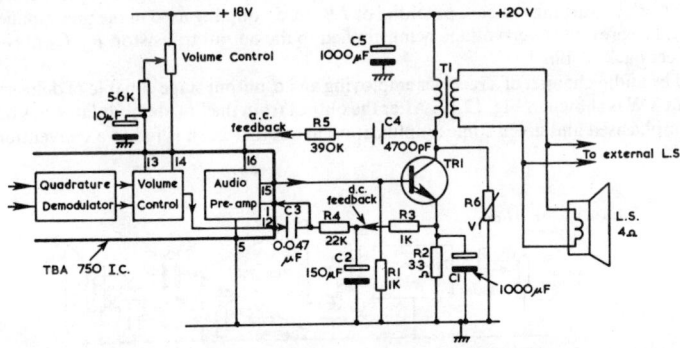

FIG. 12.11 AUDIO STAGES EMPLOYING I.C. PREAMPLIFIER AND DISCRETE COMPONENT OUTPUT STAGE

may be used after the demodulator to vary the quantity of audio signal supplied to the audio amplifiers as is normal practice. In some i.c.s, however, an electronic volume controlled stage is used as shown where the gain of the stage is controlled by an external variable d.c. voltage. A basic circuit of such a stage is given in Fig. 12.12. TR_1 and TR_2 form an unbalanced long-tail pair fed with push-pull audio inputs. This arrangement provides greater gain than a single transistor stage. Amplified audio is fed out from across R_L, the magnitude of which is controlled by the d.c. on the base of the constant-current transistor TR_3. By varying the d.c. on TR_3 base the current through TR_3, and hence TR_1 and TR_2, is varied. The voltage gain of the amplifier decreases as the effective common emitter load (TR_3 and R_1) increases. As TR_3 base is made less

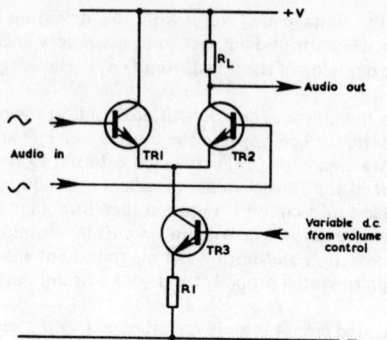

FIG. 12.12   INTEGRATED CIRCUIT D.C. VOLUME CONTROL

positive, the effective common-emitter load is increased and the gain decreased. The gain can be varied over a large range, say, 10,000 to 1 (80 dB) by this method.

The output from the volume control circuit of Fig. 12.11 is then fed through a.f. preamplifier stages. The volume control circuit output appears on pin 12 and is fed into the preamplifier on pin 1 *via* C_3. The preamplifier contains two emitter-follower stages and one common-emitter amplifier. Audio signal is fed out of the i.c. on pin 15 *via* an emitter-follower stage (R_1 is its emitter load) to the base of the output transistor TR_1. This transistor and its associated components form a class-A single-ended output stage with transformer coupling to a 4 Ω loudspeaker providing 2 W output. A.C negative feedback is used from the collector of TR_1 *via* C_4 and R_5 to one of the preamplifier stages within the i.c. D.C. feedback is taken from TR_1 emitter through R_3 and R_4 to the preamplifier (pin 1). This is used to stabilise TR_1 current against temperature variations by varying the base potential of TR_1 (d.c. coupling used in the preamplifier). R_6 (v.d.r.) prevents over-voltage being applied to the output transistor. R_3, C_2 prevent a.c. feedback to pin 1.

The audio channel of a receiver employing an i.c. output stage capable of delivering about 3 W is shown in Fig. 12.13. After the output from the f.m. demodulator has been de-emphasised and given some amplification in another i.c. it is fed to a conventional

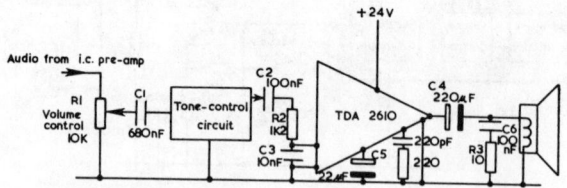

FIG. 12.13   AUDIO STAGES EMPLOYING I.C. OUTPUT STAGE (TDA 2610)

volume control R_1. The signal is then fed into a tone control circuit (which uses discrete components) and then *via* C_2 and R_2 into the output i.c. (TDA2610). This i.c. employs a class-B output circuit with a current stabiliser. The chip is encapsulated in a dual-in-line plastic package with a heat sink connector. The final output is fed to the loudspeaker *via* C_4. C_5 decouples any hum and C_6, R_3 ensure stability at the higher frequencies.

Remote Control of Sound

We will now consider how the volume may be increased or decreased remotely using ultrasonics. A system using pulse counting and dividing techniques in the ultrasonic receiver was described in connection with channel change operation and the

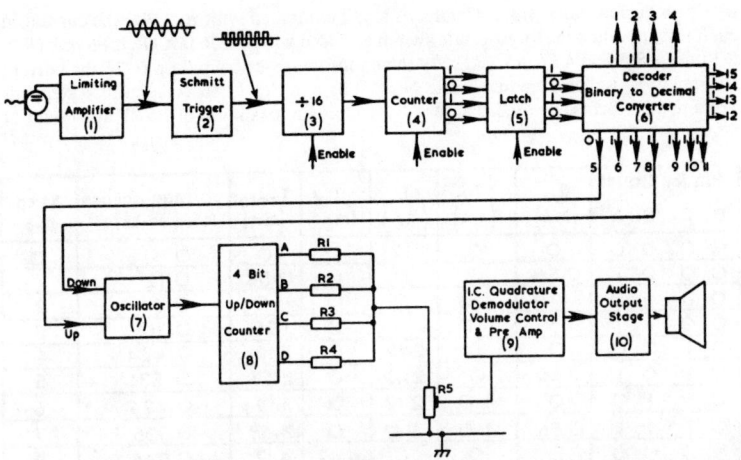

FIG. 12.14 REMOTE CONTROL OF VOLUME SHOWING OPERATION WHEN VOLUME INCREASING

same receiver system will be used to show the principle of the volume control function, Fig. 12.14.

Volume 'up' or volume 'down' is initiated by pressing separate buttons on the ultrasonic transmitter hand unit. Operating either of these buttons causes an ultrasonic wave of different frequency to be transmitted thus enabling the receiver to recognise whether volume 'up' or volume 'down' is required. At the receiver blocks (1)—(6) operate as previously described but volume 'up' and volume 'down' are represented by different binary counts corresponding to the different ultrasonic frequencies used for these functions. In block (6) the binary count is converted into a decimal output. Suppose that on operating the volume 'up' button on the transmitter the resulting binary count output of the latch, block (5), is 0101 corresponding to decimal 5. Pin 5 of block (6) goes to a low level (represented by 0) and this trigger signal is fed to block (7) (a low frequency oscillator -4 Hz). The trigger signal causes block (7) to commence oscillations. Negative-going edges of the oscillator output waveform are supplied as clock pulses to a 4-bit up/down counter which counts the clock pulses up from 0 (0000) to 15 (1111) or down from 15 (1111) to 0 (0000) using a series of flip-flop circuits. As the count increases, the outputs A, B, C and D change state in binary code with a 1 output representing a particular d.c. voltage and a 0 representing zero voltage. The presence of a 1 on any output will cause a current to flow in the appropriate resistor (R_1—R_4) and R_5. With an increasing count, the voltage across R_5 builds up in the form of a stepped-voltage waveform.

The digital output of block (8) is converted into an analogue type signal and Fig. 12.15 shows this in more detail. Here the binary coded output of block (8) is represented by switches. When any output is at binary 0 the appropriate switch is open and when at binary 1 the switch is closed. With a 10 V supply and neglecting the effect

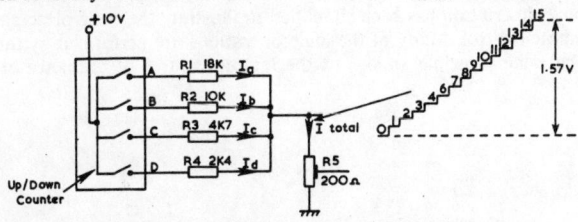

FIG. 12.15 DIGITAL-TO-ANALOGUE CONVERSION

of R_5 on current flow (since its value is small compared with R_1—R_4) the current in each resistor when the appropriate switch is closed will be 0·56 mA (R_1), 1·0 mA (R_2), 2·12 mA (R_3) and 4·17 mA (R_4). As the count increases from 0 up to 15 the current flowing in R_5 is increased in discrete steps, shown in Fig. 12.16. As a result, a stepped-voltage waveform is produced across R_5 reaching a maximum of 1·57 V at a count of 15.

Binary Count D C B A	I_a (mA)	I_b (mA)	I_c (mA)	I_d (mA)	I total (mA)	Voltage across R5 (volts)	Step No
0 0 0 0	0	0	0	0	0	0	0
0 0 0 1	0·56	0	0	0	0·56	0·12	1
0 0 1 0	0	1·0	0	0	1·0	0·2	2
0 0 1 1	0·56	1·0	0	0	1·56	0·312	3
0 1 0 0	0	0	2·12	0	2·12	0·424	4
0 1 0 1	0·56	0	2·12	0	2·68	0·536	5
0 1 1 0	0	1·0	2·12	0	3·12	0·624	6
0 1 1 1	0·56	1·0	2·12	0	3·68	0·736	7
1 0 0 0	0	0	0	4·17	4·17	0·834	8
1 0 0 1	0·56	0	0	4·17	4·73	0·946	9
1 0 1 0	0	1·0	0	4·17	5·17	1·034	10
1 0 1 1	0·56	1·0	0	4·17	5·73	1·146	11
1 1 0 0	0	0	2·12	4·17	6·29	1·258	12
1 1 0 1	0·56	0	2·12	4·17	6·85	1·360	13
1 1 1 0	0	1·0	2·12	4·17	7·29	1·458	14
1 1 1 1	0·56	1·0	2·12	4·17	7·85	1·57	15

FIG. 12.16   TABLE SHOWING BUILD UP OF VOLTAGE ACROSS R_5 FOR INCREASING BINARY COUNT (VOLUME INCREASING)

The voltage analogue of the count is then fed from R_5 slider to the volume control circuit inside an i.c. represented by block (9). Here the rising d.c. potential increases the gain of the a.f. section of the i.c. in the same way as was described for electronic volume control.

On reaching a maximum count of 15 (1111) the counter does not reset to 0 (0000) as it is internally connected to remain in the maximum count position, *i.e.* at maximum volume. If the volume 'down' button on the ultrasonic transmitter is activated a new radiated frequency is selected which is processed by the receiver to give a trigger signal on pin 8 of block (6), *i.e.* pin 8 goes to logic 0 and the other pins are at logic 1. The trigger signal restarts the oscillation in block (7) and the negative-going edges provide the clock pulses for the counter which changes from an 'up' count to a 'down' count. As the binary coded output of block (8) falls from 15 towards 0, so does the current flowing in R_5 and hence the voltage across it. In consequence, a voltage of diminishing amplitude is fed to block (9) which reduces the volume. When the remote operation facility is not being used, touch-buttons and associated circuits may be employed to provide the trigger inputs to block (7) to turn the volume 'up' or 'down'.

The above description has been simplified to illustrate the basic principle of this form of remote control. Many of the logic operations are performed by integrated circuits. The same principle applies to the remote control of brilliance or colour saturation.

INDEX